Enjoy Your Science Meeting!

A practical guide to getting the most out of attending scientific conferences

Other World Scientific Titles by the Author

Enjoy Writing Your Science Thesis or Dissertation!: A Step-by-Step Guide to Planning and Writing a Thesis or Dissertation for Undergraduate and Graduate Science Students
Second Edition
by Elizabeth Fisher and Richard Thompson
ISBN: 978-1-78326-420-9
ISBN: 978-1-78326-421-6 (pbk)

Physics with Trapped Charged Particles: Lectures from the Les Houches Winter School
edited by Martina Knoop, Niels Madsen and Richard Thompson
ISBN: 978-1-78326-404-9
ISBN: 978-1-78326-405-6 (pbk)

Trapped Charged Particles: A Graduate Textbook with Problems and Solutions
edited by Martina Knoop, Niels Madsen and Richard Thompson
ISBN: 978-1-78634-011-5
ISBN: 978-1-78634-012-2 (pbk)

Choosing a Meeting
Preparing an Abstract
Preparing a Poster
Preparing a Formal Talk
Figures and Tables
Presenting a Poster
Giving a Talk
Chairing a Session
Practical Arrangements

Enjoy Your Science Meeting!

A practical guide to getting the most out of attending scientific conferences

Elizabeth Fisher
University College London, UK

Richard Thompson
Imperial College London, UK

World Scientific

NEW JERSEY · LONDON · SINGAPORE · BEIJING · SHANGHAI · HONG KONG · TAIPEI · CHENNAI · TOKYO

Published by

World Scientific Publishing Europe Ltd.

57 Shelton Street, Covent Garden, London WC2H 9HE

Head office: 5 Toh Tuck Link, Singapore 596224

USA office: 27 Warren Street, Suite 401-402, Hackensack, NJ 07601

Library of Congress Cataloging-in-Publication Data
Names: Fisher, Elizabeth M. (Professor of neuroscience), author. | Thompson, Richard
    (Richard Charles), 1955–   author.
Title: Enjoy your science meeting! : a practical guide to getting the most out of attending
    scientific conferences / by Elizabeth Fisher (University College London, UK),
    Richard Thompson (Imperial College London, UK).
Description: New Jersey : World Scientific, 2019.
Identifiers: LCCN 2019013756| ISBN 9781786347220 (hc) | ISBN 9781786347350 (pbk)
Subjects: LCSH: Science--Congresses--Planning. | Meetings--Planning.
Classification: LCC Q101 .T5357 2019 | DDC 500--dc23
LC record available at https://lccn.loc.gov/2019013756

British Library Cataloguing-in-Publication Data
A catalogue record for this book is available from the British Library.

For any available supplementary material, please visit
https://www.worldscientific.com/worldscibooks/10.1142/Q0214#t=suppl

Desk Editors: Dipasri Sardar/Jennifer Brough/Shi Ying Koe

Typeset by Stallion Press
Email: enquiries@stallionpress.com

Printed in Singapore

Preface

When our book *Enjoy Writing Your Science Thesis or Dissertation!* was published in 2014 we discussed the other key activities that new scientific researchers engage in at the start of their careers. Writing a thesis is one of the first milestones but other important steps that a developing scientist needs to take include writing papers to communicate their science with other people in their field and going to conferences to interact with other researchers. We realised that what happens at a conference can be crucial in establishing a presence in the field. However, when we were starting out we didn't have a guide as to what meetings are all about (communication, networking, science) and how to prepare for them or what to do when going to a meeting. And no one reassured us that plenty of quite shy people go to meetings and survive (and actually quite enjoy them). So we wrote this book, for you, as a guide, based on our many decades of experience. This is what we wish we'd known when starting out.

About the Authors

Elizabeth Fisher is Professor of Neurogenetics at University College London (UCL). She was an undergraduate in Physiological Sciences at the University of Oxford, took on a PhD at Imperial College/MRC Harwell and a postdoc at MIT before coming back to Imperial College to run a lab for 11 years and ultimately moving to UCL where she has been a research Professor since 2001, now also with a lab at MRC Harwell. Along the way, she has had very, very brief careers in a variety of places including an agency for playwrights, the wine department of a major auction house, the Attorney General's office in Canberra, and a fish and chip shop. Her research interests are the creation and analysis of models of neurodegenerative disease. She is rarely seen at breakfast at meetings.

Richard Thompson is Professor of Experimental Physics at Imperial College London. He started his career as a physics undergraduate at the University of Oxford, where he also gained his DPhil. After a period as a postdoc in Germany he took up a position at the National Physical Laboratory in Teddington, UK. He moved to Imperial College London in 1986 and has been there ever since, despite the fact that he had vowed never to commute into central London. He has held a number of administrative positions at Imperial, including Director of Undergraduate Studies for the Physics Department and the Natural Sciences Faculty Senior Tutor. His research concerns spectroscopy and quantum optics with laser-cooled trapped ions.

Acknowledgements

We would like to thank our editors at World Scientific Publishing for their help while we have been writing this book, in particular Mary Simpson and Jennifer Brough. They have been extremely patient whenever we have not been able to meet deadlines (oh dear ...) and we are grateful for their understanding.

We thank Zara Slattery for the wonderful illustrations. These follow on from her work in that most excellent book *Enjoy Writing Your Science Thesis or Dissertation!* (Imperial College Press/World Scientific, 2014), and again show Professor Karloff and others in ways that perfectly demonstrate points in the book.

Richard would like to thank his wife Margaret for her support throughout the writing of this book, especially when there were conflicting priorities. Often his time spent writing about conferences displaced other important activities and he is very grateful for Margaret's tolerance and good humour when this happened. Richard has enjoyed going to many conferences over the years, with an enormous range of styles from small workshops in isolated and extremely basic facilities to huge commercial meetings. All of them have been valuable, but in different ways. He feels guilty that he has reaped the benefit of other people's hard work organising conferences but has not really done his fair share of this important activity (see Chapter 14).

Lizzy would like to thank English graduates Jim Ross for his suggestions on the Black Arts of Meetings, and Rohit Jaggi, journalist, psychotherapist, party-thrower, for his advice on word usage and abusage. She also thanks all the people who have been through her lab, and taught her so much, and the many lovely people encountered at meetings, some of whom have become lifelong friends, whom it's always a pleasure to see at the next meeting. Unlike Richard, she has organised or co-organised several meetings, and thanks all those whom she has worked with.

Contents

Crest of the University of West Cheam (which may, or may not, exist), where the postgraduate students and academics depicted in the illustrations throughout this book are based. The university, formerly the West Cheam Institute of Technologickal Arts, boasts some unconventional academic staff including Professor Karloff, Dean of Alternative Physics, and it carries out research on exotic subjects such as the behaviour of landfish. Students at the university have encountered many problems, such as failed experiments, crashed computers, lost references, broken printers and errant supervisors. At conferences some were anxious about meeting other researchers; others brought the wrong sized posters; and a few didn't practise their talks. However, despite this, they all enjoyed attending their conferences and learnt a lot from the experience!

Chapter 1

Overview

1.1 Introduction

If you are applying for a meeting or getting ready to go to one, then you probably don't want to spend time reading this book from cover to cover (although, obviously, that would be a totally enjoyable use of your limited and valuable time, we think). So this chapter is a short guide to what is in the rest of the book, summarised under the same chapter headings. Read through this chapter to select what's useful in the book for your circumstances and then go to the relevant chapter for more detail.

And, as much as possible, enjoy yourself. Meetings can be tough, but they're also the places/events where you can make lifelong friends and get really good ideas for your research. Remember that no matter how junior or senior you are, the conference is designed to be of benefit to you as much as anyone else. If you are a bit shy, remember that many others are shy too, but most of the people at the conference will be keen to make new friends and find out about their research. So to get the most out of your conference, it's worth making the effort to meet other people and learn about new scientific topics (Figure 1.1).

1.2 Chapter 2: Choosing a Meeting

Meetings come in all shapes and sizes, from three-person lab meetings to huge symposia involving tens of thousands of attendees. They all have certain features in common including the exchange of information

Figure 1.1   You've arrived! Don't worry if you feel a bit lost at first — you'll soon start to meet other people.

and interaction with other people. Here we talk about the main types of meeting that you will come across, regardless of your scientific field, and what to expect when you attend such meetings. The list includes:

- **Symposia:** Huge meetings with typically thousands of delegates.
- **Conferences:** A general term for medium-sized meetings of up to few hundred people, generally more focused on a particular area than symposia.

- **Workshops:** Smaller meetings with a narrow focus, maybe looking for solutions to a particular research issue.
- **Lab or group meetings:** These are more local meetings of a research group or a collaboration aimed at keeping everyone informed of day to day progress and, ideally, getting helpful advice from labmates.
- **Journal clubs:** Again, typically a research group getting together to try to understand recent research papers in their field.
- **Retreats:** These are often meetings away from the lab used by an institute or department to help foster collaboration and decide on research strategy.
- **Summer schools:** Usually aimed at new entrants in a field (PhD students or new postdocs) to help them get started by learning the basics from experts.
- **Commercial conferences:** Nowadays there are many organisations setting up conferences that are primarily designed to make money out of you: watch out for these! Sometimes their scientific value is very limited.

The most important point is that you engage with the meeting, and get something out of it for you, however nervous or unconfident you feel, or however much of a newbie you are.

Go ahead and register for your chosen meeting and we'll show you how to prepare for it, and then take you through the process of attending.

1.3 Chapter 3: Preparing a Conference Abstract

An Abstract is a short description of your project, often only a couple of paragraphs or so, which you may be asked for by the conference organisers. They may decide whether to take you based on your Abstract, and they may put your Abstract in the conference booklet, so it could be online forever. Also, they might decide whether to offer you a talk or a poster based on your Abstract. So Your Abstract advertises You and Your Research. It's a sales pitch! Therefore, it has to be well-written to communicate clearly what you are doing, why it is important, how you are doing it, and what you've achieved. And then again, why this research is important.

The main components of the Abstract are as follows:

- **Title:** Think of a short descriptive title that sums up your work.
- **Authors:** List the names of all the authors and their affiliations.
- **Topic:** Explain what the subject of your work is and why it's important.
- **Aim:** State clearly what the aim of your particular study is.
- **Method:** State how you carried out your experiments or calculations.
- **Results:** Summarise what you found out so far in this research.
- **Discussion:** Explain what your results mean, and how they should be interpreted.
- **Conclusions:** Summarise the main outcomes of your work and the next steps.
- **References:** Include these so that your reader can find out more if they want to.
- **Keywords:** You may also be asked for a few keywords to describe your work.

We give some guidelines for writing the Abstract, oh, and a reminder, make sure all your co-authors have agreed to the Abstract before you submit it with their names on. It is *really irritating* to be doing an online search and finding your own name on an Abstract you didn't know about!

Final word, check the Abstract deadline, because for big conferences these can be many months before the conference itself.

1.4 Chapter 4: Preparing a Poster

Posters really are just ... posters ... that advertise your work. They tend to be a standard size, usually A0 or A1 (i.e. roughly 120×85 cm or 85×60 cm), and like any other advertising poster, they should be an eye-catching array of easy-to-read and -understand information. That means use colour! Big fonts! Make it enticing! Posters are for attracting people to your work.

We give some basic rules for preparing your poster but the first golden rule is *keep it big, bold and beautiful (well, and colourful too)*. The second golden rule is *don't put in too much stuff.* In terms of content, try as hard as you can to make it easy for people to grasp hold of what you are doing. Include a short summary or Abstract so that someone who is only interested in the very broad picture knows what to look at. Make sure that there is a logical flow of the material through the poster. Look around your institute at what other people do for their posters and see what works well and what doesn't.

And also, remember the practicalities: what size should the poster be, what are the deadlines for printing, and how are you going to transport your poster?

1.5 Chapter 5: Preparing a Formal Talk

Most of us are not natural speakers, but the good news is that you can learn how to be a good speaker, even a great speaker, with enough practise and providing you do the most important thing:

prepare, **prepare**, prepare!

And that means: finding out how long you have to talk for, planning how many slides you should have (no more than one per minute of speech, on average), thinking about what you are trying to communicate and then drafting your talk, based on: **Why? How? What? Why? Who?**

- **Why** are you doing this research?
- **How** are you going about it?
- **What** have you found out?
- **Why** are your results important?
- **Who** gave you the resources to do it?

Be clear about the key points and then plan your talk in a logical order to present these points, reiterating why they are important.

When designing your slides you need to keep things as simple as possible, otherwise you distract the audience from the points you're trying to make. So keep animation to a minimum. Think about the background to your slides (why do you have a picture of the Mona Lisa for your slide background?). Use colour to maximise impact. And try not to put many words on a slide because most people can't read words, listen to what you're saying and process that all at the same time. Make good use of figures: they break up the text and illustrate the key ideas you are talking about.

In Chapter 5, we go into detail about designing slides and how to make your presentation. We also stress how important it is to rehearse your talk in a room as similar as possible to the one in which you will be speaking. Load the slides and stand at the back of the room and check that they're still visible and the fonts are large enough.

1.6 Chapter 6: Figures and Tables

People have been saying 'A picture is worth a thousand words' (and similar variants) for over 100 years. It's true! You can often use a figure to get information across much more effectively than if you had to use text alone. So it's good to use figures of different sorts freely in your posters and presentations. Tables are also very good if they are used carefully, but watch out for the temptation to include too many pieces of information in a table.

All figures and tables need to be clear and simple because your audience just is not going to have time to go through them in detail, so think carefully about presentation and how you can highlight key points. The types of figures you use in a published paper may not work so well in a poster or spoken presentation (for example, they may be too complex or the labels may be too small). In this chapter, we consider photographs, diagrams, line drawings, graphs and other plots, and provide some guidelines for how to present them.

Some key points about figures and tables that we discuss include the following:

- Use colour as much as possible, to make the points you wish to get across to your audience.

- Keep figures as simple as you can, as people don't have long to look at them.
- Keep tables to the minimum number of columns and rows necessary because it's really hard to take in a lot of information at once.
- Make sure the lettering (axis titles, data labels, titles, etc.) is all big enough to read from a distance.
- If you want to draw attention to some particular detail, indicate it clearly.

1.7 Chapter 7: Presenting a Poster

When you come to a meeting with your poster, you put it up on the poster board (usually drawing pins are provided) and then at some point in the meeting all the poster lead authors (i.e. you!) in your session will be asked to stand by their posters (the Poster Session) so that other people can stroll around and look at the posters, asking the

Figure 1.2   Young Isaac Newton waits nervously at his poster session.

authors questions. You will have different sorts of visitors: those who are really really interested and want more information, those who are casual visitors, with whom you can spend an enjoyable few minutes explaining what you do, and then shy people on their own who are walking around and would really appreciate you talking to them. Occasionally — and it's happened to everyone — you don't get anyone coming up to you in a poster session. Don't be put off by this, it's just like that sometimes (Figure 1.2).

It's a good idea to have prepared a short 'elevator pitch' of say 30 seconds to introduce your poster and summarise the main results concisely. This will be useful for people who are interested but are not experts. When you prepare this pitch, try to see it from the other person's perspective so you don't assume too much previous knowledge.

If you can, get some time away from your poster so that you can take a look at the other posters in your session. If you don't have a colleague who can look after your poster while you are away, leave a note on it saying when you will be back, and make sure you really are back at that time.

1.8 Chapter 8: Giving a Talk

The one point we stress here that is more important than anything else is to:

practise! Practise! PRACTISE!

And keep practising until you are so confident and fluent you feel like you are a supreme being in the universe. At least for your talk. And when we say 'PRACTISE' we mean 'SAY THE WORDS OUT LOUD, WITH THE SLIDES, EVEN IF YOU ARE TALKING TO YOURSELF IN YOUR BEDROOM'. There are no shortcuts for this. You may feel embarrassed, but it's much better to be embarrassed on your own than in front of 200 people when you can't find a good way of expressing something clearly.

Keep rehearsing — for one of us, when we're really nervous, we simply start by reading off the first slide … what our name is … it's a good way to start.

Come the day of the talk make sure you go to the room before-hand and check the audio-visual (AV) facilities. How does it work? Will someone be there to help you? Do you need to load your slides before you speak? See if during a coffee break you can operate the controls and bring up your slides, just to see what it feels like. Is there a laser pointer available if you need one?

When the talk finally comes, make sure you always know your first sentence, even if it's simply to say who you are and where you are from. While you are doing this, your brain can get itself into gear for the rest of the talk. Then as much as you can, look at the audience. Don't forget to do this, because they want to see you — and actually they want you to do well. Almost all audiences are supportive, so you simply have to speak and make eye contact. That's all …

In the highly unlikely event that the building catches fire, the projector blows up, or a meteorite hits the auditorium half way through your talk, then look to your Chair for guidance and do what they say. This includes getting away from the flames.

1.9 Chapter 9: Chairing a Session

As the Chair, you are the most powerful person in the room. Which can be very daunting, particularly if you have a speaker who thinks *they* are the most powerful person in the room. But audiences (and most speakers) want strong and decisive Chairs. So if you work to a few simple guidelines you should be all right.

And the first guideline is to try very hard to keep to time. Sometimes organisers don't allow enough time for interesting discussions, and the questions keep going, in which case you might be able to overrun by a couple of minutes (ONLY) and you can then stop the question session by asking people to keep the discussion going over coffee/lunch/dinner/in the bar.

Keep a copy of the meeting programme to hand, so you can introduce speakers and their topics and you don't have to worry about forgetting anyone, or the time they have to speak.

Try to find your speakers in the break before the session, tell them you are their Chair, remind them how much time they have, and then tell them how you're going to signal a 3- or 5-minute warning to them that the talk has to end. Tell them how much time they have for questions and decide whether you will field the questions for them, or they will field the questions.

At the end of your session, thank everyone for talking, and the audience for listening.

1.10 Chapter 10: Talking to the Public

We all need to talk to members of the public about what we do and usually the public is really interested and often very knowledgeable. Also, it is society that pays for us to do scientific research so it is right and proper that members of society have the opportunity to find out about what scientists are doing. Therefore 'Science and Society' activities are very important for all practising scientists. However, not everyone has a science background and so you need to tailor your talk to include a lot more explanation for this audience and absolutely no terms that a non-scientist wouldn't know. It's important that this is seen as an *engagement* with members of the public, and not just a one-way flow of information from the expert to the interested layman. So be prepared for lots of questions and suggestions.

Again, if you are giving a talk, start with the same structure:

- **Why** are you doing this research?
- **How** are you going about it?
- **What** have you found out?
- **Why** are your results important?
- **Who** gave you the resources to do it?

And think carefully about who your audience is. Of course stick to the rules for all talks, for example being aware how much time you have and how many slides you need (one slide per minute of talk is the rough rule). No equations, and very few complicated graphs please, as often members of the public don't feel confident with them. But it's good to use clear visual material where possible.

Be sensitive to your audience, particularly if you are talking about controversial topics, or medical topics. Be responsible and respectful of other people's views.

1.11 Chapter 11: Practical Arrangements for Going to Meetings

Well in advance of the meeting, check registration dates and timing, pay the fees if necessary, sort out accommodation arrangements and travel plans (including any necessary visas), and work out how you're going to pay for it all (write a travel grant application to a learned society or other funder if you have to).

During the conference, just enjoy yourself! Take advantage of the opportunity to mix with lots of other people who are interested in the same things as you are. Conferences are great for social interaction as well as for scientific communication so try to make the best of it.

After the conference, don't forget to follow-up on those new social and scientific contacts you made at the meeting. And make sure you apply for refunds of your expenses promptly, as most Universities/ meetings won't pay expenses after a 3- or 6-month cut-off from the date of the meeting.

1.12 Chapter 12: Making the Most of Your Meeting

Conferences can be a bit intimidating if you are shy, but you shouldn't worry — many other people will be feeling shy as well but will also want to make new contacts at the conference. If you make the effort to talk to other people, you will find that it's easy to get to know them and to have a good time socially. Then you will also find that you start to pick up ideas and suggestions from them that will help your research (by the way, they will appreciate good ideas from you as well).

The time at your conference will go very quickly so make sure you make good use of the opportunities. Plan in advance which talks and posters you want to see. If there are particular people you want to talk to, don't leave it too late. Go to talks to find out about new topics that might be interesting. But don't forget to take a look at the location as well if you get a chance: conferences are often in vibrant cities or beautiful locations that you might not otherwise get a chance to see.

Finally, don't wear yourself out. Attending a conference can be exhausting so don't forget to take a break from time to time so you don't burn out.

1.13 Chapter 13: Publishing Your Conference Paper

Conferences are often associated with special issues of journals or a book in a conference series or a stand-alone book of proceedings from your conference. In this chapter, we discuss the pros and cons of publishing in conference proceedings. The decision will depend very much on your circumstances, but the sort of things you should consider are as follows:

- Is there going to be some collection of proceedings from the conference?
- Will the contributions be refereed?
- Do you have material ready for publication?
- Is it better to wait till you have more substantial results?
- Will the proceedings appear in an established journal?
- Is it the right place for your results?
- Have you got time to write an additional publication?

Sometimes a conference paper can be a useful additional publication, but in other cases it can just sit on people's shelves gathering dust. So weigh up the pros and cons carefully before committing to preparing a conference paper.

1.14 Chapter 14: Organising Your Own Meeting

Before long, someone will ask you to organise a meeting. You can avoid this for a while but eventually you will run out of excuses. And in fact, this job can be quite fun. Especially if you like hosting parties.

In this chapter we discuss some of the things that you need to do if you are organising a small to medium size meeting, with or without a budget. There are a number of points you should think about:

- Get a reliable team together — it's too much to do this on your own!
- Decide on the subject for the meeting and choose a suitable title.
- Decide on a date but remember that you may have to be flexible if there are particular people you want to attract.
- Contact your keynote speakers as soon as possible.
- Find a suitable venue.
- Draw up your timetable. Not as easy as you might think, but it's worth putting in the effort to get this right.
- Sort out all your other speakers.
- Advertise your meeting as widely as possible and get people to tell you if they are coming.
- See if you can get some funds together so you can at least provide teas and coffees to your delegates.
- On the day, get there early, be prepared for problems but otherwise enjoy yourself!

For larger meetings, you will probably need to work with a larger team and make use of professional conference organisers. They are the experts and will know what to do.

Organising a meeting is a lot of work so you don't want to go into it without considering how much effort it is. But it can be very rewarding and your efforts will be appreciated by your colleagues.

Key Points

- Think about what you want to get out of a meeting and then choose the one that will be most productive for you.
- Check the Abstract deadline and prepare your Abstract in good time, writing it as clearly and concisely as possible.
- Before you start preparing a poster, look at posters around your department to see what works well.

- Make your poster visually attractive: make good use of figures and don't use too many words.
- Get your poster ready in good time to allow for printing.
- If you are speaking, find out how long you have to talk and tell your story in a logical, not chronological, order.
- Keep the slides as simple as possible, making good use of suitable graphics.
- At the meeting, make an effort to be friendly to other people and enjoy their company.
- Prepare a 30-second elevator pitch before the poster session and don't assume that your visitors already know a lot about what you are doing.
- Practise your talk to the correct timing.
- Check that your talk works on the conference computer and AV system.
- When talking, keep calm and look at your audience as much as possible.
- If you are chairing a session, try hard to be firm with your speakers and keep them to time.
- When talking to the public, concentrate on simple concepts well explained, using analogy as appropriate.
- Be sensitive to your audience and try to enter into a genuine two-way conversation with them.
- After the meeting, do your travel expenses immediately and follow up on contacts you made.
- If there will be conference proceedings of some sort, consider carefully whether to submit an article or not.
- If you are asked, don't underestimate how much time it will take you to organise a meeting yourself!
- If you do agree to organise a meeting, get a budget if you can and get other colleagues to help you.

Chapter 2

Choosing a Meeting

2.1 Introduction

Meetings vary in size. Some of the really large biomedical meetings that take place in the USA, for example, can have over 30,000 attendees. Your lab meeting might have three people. Regardless of size, what everyone is doing is interacting with other people in the field, and presenting their data and themselves. Here we give a brief guide to the sorts of meetings available to us academics. There is some overlap in the titles — for example, a 'meeting' and a 'conference' are basically the same thing — so if you are interested in attending, always look at the meeting website (if there is one) and see if you can find out more information from your experienced colleagues.

Think about what you want to get out of attending a conference before deciding which one to attend. Is the main point to present your research and make sure that your latest results get recognised in the community? Or is it because you want to find out what other people are doing and catch up on the latest developments? Do you want to get to know some of the experts better and pick their brains about difficulties you have been having with calculations or experiments? Maybe you want to try to establish new collaborations. Or maybe you want to see if there are any jobs going in your field and to check out possible supervisors.

Choosing which conferences to go to is important. You have very limited time and usually very limited cash, so first find out who is talking to see if the meeting really is relevant and helpful — and get advice from your supervisor or line manager, and other people in your field.

2.2 The Main Types of Scientific Meetings

The different meetings that scientists organise serve different purposes. Let's start large and then get small.

Symposia

The biggest meetings tend to be the large international symposia. The word 'symposium' is derived from an ancient Greek word that essentially means a drinking party ... and this fine tradition is usually upheld enthusiastically by modern scientists. A symposium is a big get-together of experts and interested people, to discuss a broad topic, which today could be anything from atmospheric physics to landfish biology. By 'big' we mean anything from a few hundred to perhaps many thousands of people. This type of large meeting may have many parallel sessions running at the same time because there are too many talks for the time available, and often no rooms that will take the entire group of attendees at the symposium.

Large meetings like this are often associated with a major exhibition from companies providing services, equipment or publications in that field. This can be a big attraction of a symposium if you want to find out about new products that may be relevant to your research. It will give an opportunity to see new equipment in action — always better than just reading a description — and you can talk to the representatives from the supplier as well. You can ask them detailed questions about their products and you may be able to negotiate a deal with them. As well as equipment there may be stands from suppliers of commercial research services and publishers where you can look through recent books and journals to see if they are useful to you. And if you do go to an exhibition, don't forget to stock up on pens, pads of paper, USB sticks and whatever else they are handing out as freebies!

Conferences

To confer is to consult with one another, hence we get together in conferences to do exactly that. Conferences tend to be from one hundred to a few hundred people (but can be a lot more), and are more specialised than symposia. They provide an opportunity to find out about progress in your own field but also in related areas — maybe you can pick up some ideas for techniques/theories/hypotheses that could be applied in your own work. If you are thinking about altering the direction of your research, attending a conference is a great way of informing yourself of other related areas where your expertise may be applicable, and where many of the words and concepts will be familiar.

Workshops

These tend to be smaller meetings from 50 to a couple of hundred people, depending on the field. They are usually fairly specialised within one research topic. Workshops are often a very good way to get to know the experts and what is the state-of-the-art in your field. As these meetings are smaller, and full of like-minded people, they're often friendly and good places to go if you're starting out, or not feeling super confident. They give lots of opportunities for detailed discussions with other researchers and exchange of ideas. A sign of a good workshop for you is if you look at the list of speakers and find that there aren't any people you don't want to hear.

Meetings

We've included the generic term 'meetings' here — and really a meeting can be any size, although they usually range from tiny (a three-person lab meeting) to medium (a few hundred people) and not the massive size of many symposia. However, one of the largest 'meetings' we know of takes place in the USA yearly, and is on the nervous system and regularly has more than 30,000 attendees. So always check the meeting website for further information, so that you know what to expect.

Lab or group meetings

Lab meetings are literally just that: a meeting of the people in your laboratory or research group, to present your data, usually warts and all, to each other. In the best possible circumstances, these meetings should give you support in the face of experiments not working, or your calculations reaching a dead end. They should also give you good ideas for how to troubleshoot and to write up results. Occasionally collaborators might join in to review joint progress on research (Figure 2.1).

Most lab or group meetings fall into one of two styles, and both are useful:

(1) Meetings in which you present your raw data, straight from the lab or computer, usually still in your lab book, to your labmates, for their comment and constructive criticism or helpful suggestions (ideally): This is valuable for tackling the day to day problems of undertaking science, and getting suggestions for troubleshooting techniques from experienced co-workers.

Figure 2.1   Lab meetings may just be three people (where are the biscuits?).

(2) Meetings in which you present a more polished talk, with slides, about the progress of your project, which may, or may not, show raw data: This is valuable for giving you practise in how to make formal presentations, particularly if you get little chance for this other than in fairly rare external meetings.

Both styles of group meeting are a good experience for scientists.

One of the authors couldn't decide in which style to hold their lab meetings (1) raw data or (2) formal presentation, and so now their lab meeting alternates weekly between the two types of presentation, which gives everyone a chance to present their data in the different ways. Always with biscuits of course.

Journal clubs

A Journal club is a specific type of group meeting where you discuss a particular recent paper from another group instead of discussing recent progress in your own group. This is a good way of keeping up to speed with the research field. Usually one person goes through the paper in detail in advance and then presents it to the other people present (who ideally will have at least read it through beforehand!). Apart from learning about new techniques and new developments, it's also a good way to gain experience of giving an informal presentation to other people.

Retreats

A retreat is an act of withdrawing. And that can be very useful for the modern scientist: a chance to get away from the lab, from the telephone, from the internet, from Facebook, and to have an opportunity to learn what our colleagues are up to, and to think. Often an individual lab, a department, or even an entire institute will organise a retreat so that the people within it simply have the time to talk to each other. These meetings may just be an 'awayday' where you can spend time together close to the institute, or they may be residential, lasting for a couple of days somewhere further away.

Retreats are extremely valuable for making new friends, and new collaborations when you discover that the woman across the corridor is working on a new technique that would help your experiments to understand mouse respiration or the chemistry of palladium.

One of the authors, when a postdoc many many years ago at a well-known technology institute in Cambridge, USA, used to go to the yearly Retreat. These were legendary events for the partying on the first night and there are many only slightly libellous tales to be told of the antics of now rather famous senior scientists. FORTUNATELY this was before the days of social media, so if you are going to end up drunk in someone else's room, try not to be photographed.

Summer schools

Summer schools are for learning about a new topic and are particularly aimed at new PhD students in a field or other scientists who are moving into that field. If you go to one as a participant, they are a fantastic way of getting a detailed overview of the field from experts. The format and location usually allows a lot of time for discussion and questions.

If you are presenting at a summer school, it gives the opportunity to take a broader approach to speaking about your field (and not just your own research). It obviously calls for a slightly different approach to preparing your talk(s) and we will discuss that later in Chapter 5.

Conferences designed to make money out of YOU

We need to mention scam conferences that are after you and your money. If you are in science for any length of time you will start to receive frequent emails inviting you to a conference, often using your first name (or a part of your email address), and inviting you personally as an honoured guest. A sign that something is wrong is when the person inviting you praises your deep knowledge of condensed matter physics, whereas in fact you work on deep space biology. Often the wording is completely vague (probably because the email is going to thousands of people) and asks you to suggest your own topic.

Sometimes, the invitations manage to ask you to a conference actually related to your field, and this may be in an exotic location that you'd rather like to visit. Next step: despite valuing your exalted presence, these conferences are likely to charge you a hefty fee to attend — because most of the time they are money-making exercises run by companies. A good tell-tale sign is that you have to pay more if you are giving a talk! Always check the names of the organising committee (if there is one) and the other invited speakers when assessing whether to attend a meeting like this.

2.3 Registering for a Meeting

All larger meetings invite attendees to register for a place. This gives the meeting organisers a chance to see who wants to come, and to be selective if the meeting is overwhelmed by people; and it means they can sort out the practicalities such as the number of lunches they need to order.

When you register, you usually go onto the meeting website and fill out an online form wanting details such as who you are and where you are from, and sometimes, if there is competition to get into the meeting, the site may require some information about what you would contribute to the meeting. They may ask you to submit an Abstract to help in making that decision. Larger meetings usually are not selective. Smaller meetings can be, especially if the topic is very popular. You will probably also be asked to pay, and depending on the meeting this payment will be for registration (to cover the administrative costs of the meeting which may include venue hire) and could also be for accommodation and for food and drink. The conference website should clearly explain what the charges are for.

If you need to find accommodation, then often the meeting organisers will have arranged a deal with local hotels so that you can stay in them more cheaply than the advertised rates. If so, make sure you get in quickly before they run out of rooms! Another possibility if many people are going to the same meeting is to get together a small group, and hire an apartment. That can be cheaper than staying in a hotel and may be more fun if you are with friends (see Chapter 11).

Key Points

- Know your meeting. Check the meeting website and talk with colleagues to find out exactly what you are going to.
- Choose the meeting that will be most productive for you.
- Is it important for you to have the opportunity to promote your own work?
- Will you get a chance to meet and get to know key researchers in your field?
- Look up the topics covered by the meeting, and the names of invited speakers.
- Think about what you want to get out of the meeting.
- And that includes lab meetings.

Chapter 3

Preparing a Conference Abstract

3.1 The Purpose of Conference Abstracts

If you want to present your work at a conference, either in a talk or in a poster, you will be asked to submit an Abstract in advance. The Abstract should give a concise summary of the research you want to present. The scientific content of the Abstract needs to be easily readable and interesting.

For some conferences the number of available slots for talks and/or posters is limited and your Abstract will be looked at by the organisers to help them decide who to invite to present their work and perhaps whether it should be as a poster or a talk. Obviously if this is the case you want to make sure that your Abstract is written clearly and conveys the importance of your work.

At most conferences the Abstracts will be printed in a conference *Book of Abstracts*, which will be circulated to all the participants — who will then use the *Book of Abstracts* to decide in advance which talks to listen to or which posters to look at. Again this means that writing a good Abstract is important for getting your work seen. So it is worth spending some time on.

The other important point about the Abstract is that the participants will take it away with them afterwards, so if they are interested in what you have done, the first place they will look for more information is the Abstract. For this reason, you need to make sure that it gives them the information they need to find out: who you are and

Figure 3.1   Abstract books can be quite large — but fortunately nowadays they are also usually available in an electronic format.

where you work, how to contact you, what your work is about and where it has been (or will be) published (Figure 3.1).

3.2 Writing the Abstract

As the Abstracts may be going straight into a printed booklet and possibly also online, almost all meetings request that Abstracts are written in a precise format, usually with a limit on the number of words and often with instructions about punctuation and layout.

Check if there is an Abstract template and if so, follow it carefully.

For almost all Abstracts, you will need to include a title, the co-authors, the affiliation of the authors, and finally the scientific content and possibly a few key references. You need to get the agreement of everyone on the list to the content of the submitted Abstract, because their names, as well as yours, are on the published Abstract ... and that means you need time to circulate the Abstract to everyone and get their comments back, before you can submit it. So in order to give your co-authors a few days in which to read your Abstract, you have to work well in advance. You should never submit an Abstract (or any other work) without the prior consent of the people you name as co-authors, even for small meetings internal to your Department or Institute. Partly this is for professional courtesy, partly because you may be writing something a co-author does not agree with, or does not want their name attached to, and partly it is absolutely infuriating if someone puts your name on a piece of work without either asking you, or letting you know it's out there in the literature (this has happened to one of us, twice, and we're *still* annoyed about it ... and to one of our colleagues who found his name mis-spelt as the co-author of an Abstract he hadn't seen prior to reading it at a meeting).

Think carefully about your title: it shouldn't be so short that it doesn't convey any information, but on the other hand it shouldn't be so long that people lose the will to live before they get to the end of it. Most people will only read the title before they decide to move on, so include enough information that the people who *should* come to your talk or poster will *realise* that they should come to it. As a bonus, if it's well written you will also attract people who may read further because it looks interesting to them even though it isn't in their particular area of research.

An Abstract should be a short piece of writing, typically with a maximum of one page, or shorter if the word limits say so, that very briefly introduces your work within the field of the meeting, then describes exactly what you are doing, then gives a brief account of your results and conclusions, and finally broadens out again to say why they are important. The key words here are 'short' and 'brief', so don't cram in as many words as you can onto one page.

You may be able to include a figure in your Abstract but check the instructions as there may well be restrictions. If you do include one, try to choose the figure that will be most useful — probably one of your main results — but make sure that it makes sense with little explanation as you won't have space in your Abstract to give any details about what's shown.

If the conference is months ahead (some conferences demand Abstracts 6 months before the conference itself) then you may have to hope you have some results by the time of the meeting. Write the Abstract simply saying what research you are doing currently and what you hope it will tell you.

Like all pieces of scientific writing, the Abstract should have a structure, even if it's not divided up into numbered sections; like most writing the structure is BROAD (place the work in context), NARROW (talk specifically about your own experiments and results), BROAD (place your work in context and even think about future work).

- **Title:** As we discussed above, make this informative without being too long.
- **Authors:** List all the authors of the Abstract and their affiliations.
- **Topic:** Start with a couple of sentences that explain what the subject of your work is and why it's important. Help people to understand the motivation for what you have done.
- **Aim:** State clearly what the aim of your particular study is.
- **Method:** State how you carried out your experiments or calculations.
- **Results:** Summarise what you found out in this research, and possibly what you hope to find out if you are still working towards the results.
- **Discussion:** Explain what your results mean, and how they should be interpreted.
- **Conclusions:** Summarise the main outcomes of your work and the next steps.
- **References:** Include a couple of key references so that your reader can find out more if they want to.

Of course if your Abstract is limited to five lines of text you won't be able to include everything listed above but still try to cover all those aspects to some extent if you can.

Example: for a meeting on Landfish Parapsychology

Title: **Investigating the teleportation ability of common landfish (*Gallos fritos*)**

Authors: Karloff, Barold; Brave, Gandalf T.; Pevensie, Lucy

Affiliation: Department of Landfish Biology, University of West Cheam, UK

Abstract

Landfish are widespread on Earth and elsewhere. The common landfish (*Gallos fritos*) has occasionally been described teleporting between Earth and Mars, where it may be attracted to amethyst-rich underground caverns, although the reason for visiting these sites is unknown. We have been studying the few descriptions of landfish teleportation to determine why this behaviour occurs and its physiological basis. We present our results on likely reasons for this phenomenon. We find landfish have an unusually high tropism for purple light, compared to the light tropisms of the related large groundfish (*Meleagris holidais*). This may partially explain their visits to Mars. The tropism for purple light may drive other aspects of landfish biology, including their frequent visits to nightclubs, and is a potentially important fundamental insight into the behaviour of these common creatures.

Keywords: common landfish, teleportation, Mars.

If we breakdown the above Abstract, which was recently submitted to the Annual Meeting of Landfish Biology, session on Parapsychology, we see the that the author, Professor Barold Karloff, has taken the correct approach. He starts broad: 'Landfish are widespread ...' so he gives an introductory sentence that we all feel comfortable with.

Then he narrows down to his topic. 'The common landfish (*Gallos fritos*) has occasionally been described teleporting' so that we know what the Abstract will be about, and this is in agreement with the title.

After this he presents his results: 'We present our results on likely reasons' ... so now we know exactly what the poster is going to tell us, and we can work out whether it's of interest to us.

And he then finishes by stating the importance of his results to the field: '... is a potentially important fundamental insight into the behaviour of these common creatures'. So now we understand why the work in this Abstract has significance.

Keywords: Professor Karloff was asked to provide up to four keywords for his Abstract. These are important as they will go into the index of the Abstract book. He has chosen what he thinks are the most important keywords to attract the interested reader.

This is a successful Abstract: it is short, tells you everything you need to know to decide if you are interested and it is written as much as possible in the active, not the passive, which cuts down on words and keeps the piece as dynamic as possible.

Just to illustrate what *not* to do, here is exactly the same Abstract written as far as possible in the passive with all the other redundant words and phrases that we shouldn't use. Note how much longer it is than Professor Karloff's original and how much more difficult to follow:

Landfish are known to be widespread both on Earth and also elsewhere as well. It has been reported in a number of papers recently and in the historic literature that the common landfish (**Gallos fritos**) has occasionally been described to teleport between Earth and the planet Mars, where it is thought landfish may be attracted to amethyst-rich caverns that are found underground, although currently at the present time the reason for the visits of landfish to these sites remains unknown. There are few observations or descriptions of landfish teleportation events to be found already published in the current literature. We have been studying these descriptions to try to determine why this behaviour may occur and in addition what might be the physiological basis of the phenomenon of teleportation. Here, we will present our current results on the likely reasons for this type of event taking place. It has been determined by us that landfish have an unusually high tropism for light that is purple in colour, compared to the light preference of the evolutionarily related large groundfish (**Meleagris holidais**), which does not have a high tropism for light that is purple. It is thought that this may partially explain the visits of landfish to the planet Mars, which we discussed above. The tropism of landfish for light of a purple colour is thought to potentially drive many other aspects of landfish biology, including the widely observed frequent visits of landfish to nightclubs, and thus our finding of this purple light tropism is actually a potentially important fundamental insight into the behaviour of these common creatures. This phenomenon will continue to be the subject of more ongoing investigations in our group going forward.

3.3 Submitting a Conference Abstract

You may also be asked to submit an Abstract when you register, or you may be able to do this later. Most of the time this will only be required if you want to make a presentation of some sort. Sometimes you can choose whether this is for a Poster or a Talk, and sometimes the organisers will make that decision. With a small and very popular meeting, the organisers may request Abstracts from attendees, especially students, in order to work out who to pick, so that they have a good range of interests represented. In this case you will not necessarily make a presentation.

The meeting will have a deadline for Abstract submission, which gives the organisers enough time to choose what they want and who they want, and to get the Abstracts printed in the conference booklet. Check the Abstract deadline very carefully if you want to submit one; these deadlines can be surprisingly early.

In reality many academics are **really bad** *at keeping to deadlines so they often miss Abstract deadlines and then conferences end up extending the deadline because they expect more papers to come in. As a result, people in some fields now* **expect** *the deadlines to be extended so they don't take them seriously any more. However, it's dangerous to assume that a deadline will be extended so always try to keep to it if you can, and if there's a problem, email the organisers so they know in advance that you will be late.*

As we mentioned above, the organisers may insist on you using a particular format for your Abstract or a specific template. They may insist on you preparing your Abstract with a particular font or even word processor program (such as Microsoft Word or LaTeX), or they may just ask you to send a PDF generated from any word processor.

Check the requirements carefully — and always use a font that is clearly readable, not like this:

Investigating the teleportation ability of common landfish (Gallos fritos).

Or this:

Investigating the teleportation ability of common landfish (Gallos fritos).

Apart from anything else, you run the risk that an obscure font will come out completely differently on a different computer.

Key Points

- Check the Abstract deadline and prepare your Abstract in good time.
- Show the Abstract to all your proposed co-authors so they can agree to be co-authors and have enough time to provide feedback.
- Follow the guidelines or template carefully.
- Choose an informative title.
- Choose the keywords that are most relevant to your Abstract.
- Write clearly and concisely.
- Proofread your Abstract carefully before finally submitting it.

Chapter **4**

Preparing a Poster

4.1 Why Do We Have Posters?

Posters are literally, that: posters, advertising your work. They are usually A0 or A1 in size (i.e. roughly 120×85 cm or 85×60 cm), and contain text and figures to explain your research.

Posters are not poor relations to giving a talk. Posters are wonderful things in their own right, because they give time, and a usually fairly relaxed time (often with wine) for people who are really interested to come over and learn what you're doing. They allow a two-way communication that is not possible with a talk. Many people prefer to give a poster than to give a talk, as it is less scary and more interactive and friendly — and sometimes more useful as you get more feedback.

Posters at conferences are a relatively new idea. As conferences became more popular, and more people wanted to present their work, it was no longer possible for everyone to give a talk. Rather than have too many parallel talks sessions, people were given the opportunity to present their work at a poster session where they would stand by a board giving key results so that they could discuss and explain their work personally to conference attendees who were interested. Originally it was not generally possible to print on large sheets of paper so many people prepared their posters with many A3 or A4 sheets of paper arranged in order on the poster board. Nowadays though it is unusual to see posters using anything except properly printed A0 or A1 sheets of good quality paper, although cloth posters are becoming more popular as they are easier to carry with you than a paper poster. A colleague of ours has tried wearing a cloth poster as a mini-cloak; it works well at parties.

*When one of us presented at their first ever poster session, when the idea was very new, they just took two copies of a recent printed journal article along, cut it up and stuck the pages in order on the poster board. Not surprisingly, no-one took any interest at all in the poster and a very important lesson was learned about how **not** to prepare for a poster session.*

We take the view that posters should be self-explanatory, that is, someone should be able to look at your poster without you being there to explain it, and be able to understand your research. They therefore include short explanations and descriptions as well as key figures and equations. This is the sort of poster that is particularly suitable for conferences where posters are continuously on display for a day or two and not just at the poster session. A poster like this is also useful once you have come back from your conference as you can put it up in your lab or office and use it to help explain your work to a visitor.

On the other hand, some colleagues think that the text is distracting. They would say that a poster is really just a substitute for what you might write on a white-board while explaining your research to a visitor in your office. They would therefore only include keywords along with the figures and equations and would include little text. These posters are not designed to be used when you are not there to give the verbal explanation. Although we agree that too much text is definitely not helpful, we feel strongly that a poster should stand on its own and should include enough explanation, description and discussion so that it can be understood without you present. This is the type of poster that we will describe in this chapter.

4.2 Designing Your Poster

Posters need to attract people. They need to be colourful and interesting. This means not writing masses of words in a very small font that people cannot read (especially more mature researchers to whom you may wish to apply for a job in the future ...). Your audience has to be able to read the poster from some distance away, especially if there are other people in the way, chatting about the work.

Remember most people only spend about 20 seconds looking at a poster before moving on. If you grab their attention then they will look in more detail and possibly talk to you, and that's when you can get real engagement and discussion going.

The basic principles to bear in mind are as follows:

(1) Keep the message simple so people can understand something about your work after looking at your poster for just a couple of minutes. You can give much more detail in discussion when someone has expressed an interest in finding out more.
(2) Make the poster visually attractive by the use of colour and by breaking the text up into manageable chunks. Try not to let the poster look cluttered.
(3) Don't be tempted to include too much text: keep it short and to the point and use a font that is big enough so that someone can read the poster from several paces away.
(4) Use figures to break up the text and to illustrate key points. It's often easier to understand something if there is a visual element to it rather than just words.
(5) Make sure your figures are clear and without clutter. Don't let the labels, data points or lines be too small.
(6) Include references so people know where to look for more information.
(7) Put your contact information on the poster (including your email address and web page) — some people even print a QR code on their poster that visitors can scan when they visit.
(8) Don't forget to acknowledge your University or Institute and also your funders! Perhaps by including their logos on the poster.

You can arrange the material on your poster in several different ways. Generally it's hard to read a poster if the text runs across the whole width of the sheet. This is especially true if your poster is in a landscape rather than portrait orientation. For this reason, many people use columns of text. Usually, this will flow in a logical sequence rather like you might arrange the sections of a research paper (e.g. Introduction, Materials and Methods, Results, Discussion, Conclusion). Alternatively, you can arrange the text in a series of boxes with borders that make it clear how it is broken up. These may be labelled in conventional sections as indicated above or they may be separated into topics — perhaps individual experiments or different models for some data. In this case it may not be necessary to read the sections of the poster in a predefined order if each topic is self-contained. Which type of style suits you and your work best is for you to decide.

You can use a range of software for the preparation of posters. Many people use PowerPoint but in the physical sciences it's quite common to use LATEX. There are other powerful graphics packages available that allow you to arrange blocks of text and graphics on the page in a very flexible way. Whichever software you use, it's worth taking the time to learn how to use it properly so you don't find yourself struggling to get everything to fit on the page at the last minute with no one available to help you out.

It's a good idea to walk around your institute or department and look at the posters on display in corridors. Look at how people use the space and try to assess what is effective and what doesn't work so well. Do they look visually attractive? When you have started to read a poster, does it draw you in further or does it get boring? When you have finished reading it, do you feel as if you have learned something or has it just left you confused?

Try to learn from other people's successes and mistakes to make your own poster as good as possible.

Figures 4.1–4.3 show some examples of good and not so good practise in designing posters.

4.3 Preparing the Text and Figures

You decided on the title of your poster when you submitted your Abstract to the conference. Only change it now if it's really necessary. Make sure the title is printed large enough so that people can read it from a long way away. You will probably want to have an Abstract or

Investigating the teleportation ability of common landfish (*Gallos fritos*)

Karloff, Barold; Brave, Gandalf T.; Pevensie, Lucy
Department of Landfish Biology, University of West Cheam, UK

Abstract

Landfish are widespread on Earth and elsewhere. The common landfish (*Gallos fritos*) has occasionally been described teleporting between Earth and Mars, where it may be attracted to amethyst-rich underground caverns. We have been studying the few descriptions of landfish teleportation to determine why this behaviour occurs and the physiological basis of teleportation. We find landfish have an unusually high tropism for purple light, compared to the light tropisms of the related large groundfish (*Meleagris holidais*). The tropism for purple light may drive other aspects of landfish biology, including their frequent visits to nightclubs, and is a potentially important fundamental insight into the behaviour of these common creatures.

Background

Et ultrices neque ornare aenean euismod elementum. Id semper risus in hendrerit gravida rutrum. Eget aliquet nibh praesent tristique magna sit amet purus. Congue quisque egestas diam in arcu cursus euismod quis viverra. Vulputate sapien nec sagittis aliquam malesuada bibendum. Arcu bibendum at varius vel. Vel risus commodo viverra maecenas accumsan. Id aliquet lectus proin nibh nisl condimentum id.

Apparatus

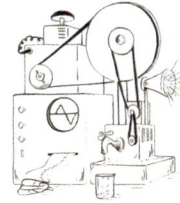

Sit amet cursus sit amet dictum sit amet. Et leo duis ut diam quam. Lobortis scelerisque fermentum dui faucibus in ornare quam viverra. Eleifend quam adipiscing vitae proin sagittis nisl rhoncus mattis rhoncus.

Theory Results

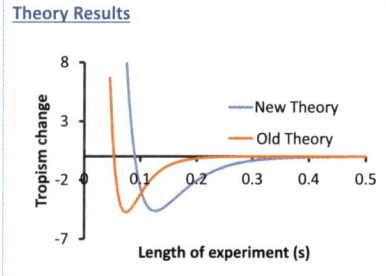

Lorem ipsum dolor sit amet, consectetur adipiscing elit, sed do eiusmod tempor incididunt ut labore et dolore magna aliqua. Ut enim ad minim veniam, quis nostrud exercitation ullamco laboris nisi ut aliquip ex ea commodo consequat.

Experimental Results

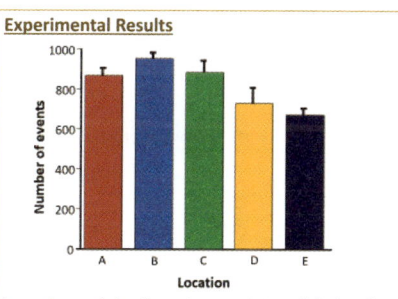

Lorem ipsum dolor sit amet, consectetur adipiscing elit, sed do eiusmod tempor incididunt ut labore et dolore magna aliqua. Ut enim ad minim veniam..

Conclusions

- Fermentum odio eu feugiat pretium nibh ipsum.
- Aliquet risus feugiat in ante metus dictum.
- Scelerisque felis imperdiet proin fermentum leo vel.
- Vestibulum lorem sed risus ultricies.

References and Acknowledgements

Mollis aliquam ut porttitor leo a diam sollicitudin tempor. Fermentum odio eu feugiat pretium nibh ipsum. Aliquet risus feugiat in ante metus dictum. Scelerisque felis imperdiet proin fermentum leo vel. Vestibulum lorem sed risus ultricies. Condimentum lacinia quis vel eros donec ac odio.

Figure 4.1   An example of a fictitious poster that is set out in a logical and readable manner, using colour effectively with boxes to separate it into logical sections.

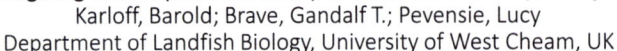

Investigating the teleportation ability of common landfish (*Gallos fritos*)
Karloff, Barold; Brave, Gandalf T.; Pevensie, Lucy
Department of Landfish Biology, University of West Cheam, UK

Abstract

Landfish are widespread on Earth and elsewhere. The common landfish (*Gallos fritos*) has occasionally been described teleporting between Earth and Mars, where it may be attracted to amethyst-rich underground caverns. We have been studying the few descriptions of landfish teleportation to determine why this behaviour occurs and the physiological basis of teleportation. We find landfish have an unusually high tropism for purple light, compared to the light tropisms of the related large groundfish (*Meleagris holidais*). The tropism for purple light may drive other aspects of landfish biology, including their frequent visits to nightclubs, and is a potentially important fundamental insight into the behaviour of these common creatures.

Background

Et ultrices neque ornare aenean euismod elementum. Id semper risus in hendrerit gravida rutrum. Eget aliquet nibh praesent tristique magna sit amet purus. Congue quisque egestas diam in arcu cursus euismod quis viverra. Vulputate sapien nec sagittis aliquam malesuada bibendum. Arcu bibendum at varius vel. Vel risus commodo viverra maecenas accumsan. Id aliquet lectus proin nibh nisl condimentum id.

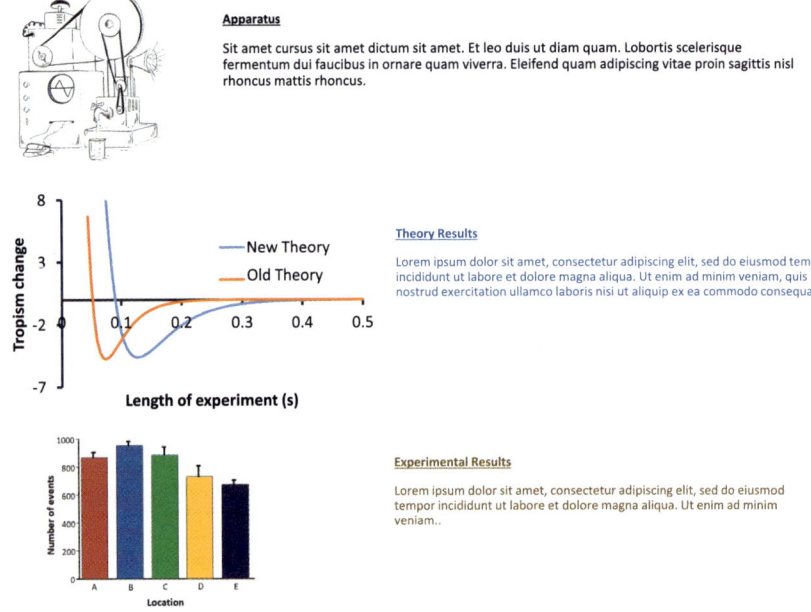

Apparatus

Sit amet cursus sit amet dictum sit amet. Et leo duis ut diam quam. Lobortis scelerisque fermentum dui faucibus in ornare quam viverra. Eleifend quam adipiscing vitae proin sagittis nisl rhoncus mattis rhoncus.

Theory Results

Lorem ipsum dolor sit amet, consectetur adipiscing elit, sed do eiusmod tempor incididunt ut labore et dolore magna aliqua. Ut enim ad minim veniam, quis nostrud exercitation ullamco laboris nisi ut aliquip ex ea commodo consequat.

Experimental Results

Lorem ipsum dolor sit amet, consectetur adipiscing elit, sed do eiusmod tempor incididunt ut labore et dolore magna aliqua. Ut enim ad minim veniam..

Conclusions

Fermentum odio eu feugiat pretium nibh ipsum. Aliquet risus feugiat in ante metus dictum. Scelerisque felis imperdiet proin fermentum leo vel. Vestibulum lorem sed risus ultricies.

References and Acknowledgements

Mollis aliquam ut porttitor leo a diam sollicitudin tempor. Fermentum odio eu feugiat pretium nibh ipsum. Aliquet risus feugiat in ante metus dictum. Scelerisque felis imperdiet proin fermentum leo vel. Vestibulum lorem sed risus ultricies. Condimentum lacinia quis vel eros donec ac odio.

Figure 4.2   The same information as Figure 4.1 but because the space is used less effectively the font size had to be reduced, making the poster less readable.

Investigating the teleportation ability of common landfish (*Gallos fritos*)

Karloff, Barold; Brave, Gandalf T.; Pevensie, Lucy
Department of Landfish Biology, University of West Cheam, UK

Abstract

Landfish are widespread on Earth and elsewhere. The common landfish (*Gallos fritos*) has occasionally been described teleporting between Earth and Mars, where it may be attracted to amethyst-rich underground caverns. We have been studying the few descriptions of landfish teleportation to determine why this behaviour occurs and the physiological basis of teleportation. We find landfish have an unusually high tropism for purple light, compared to the light tropisms of the related large groundfish (*Meleagris holidais*). The tropism for purple light may drive other aspects of landfish biology, including their frequent visits to nightclubs, and is a potentially important fundamental insight into the behaviour of these common creatures.

Background Et ultrices neque ornare aenean euismod elementum. Id semper risus in hendrerit gravida rutrum. Eget aliquet nibh praesent tristique magna sit amet purus. Congue quisque egestas diam in arcu cursus euismod quis viverra. Vulputate sapien nec sagittis aliquam malesuada bibendum. Arcu bibendum at varius vel. Vel risus commodo viverra maecenas accumsan. Id aliquet lectus proin nibh nisl condimentum id.

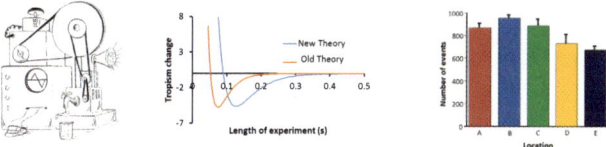

Apparatus Sit amet cursus sit amet dictum sit amet. Et leo duis ut diam quam. Lobortis scelerisque fermentum dui faucibus in ornare quam viverra. Eleifend quam adipiscing vitae proin sagittis nisl rhoncus mattis rhoncus. Suspendisse potenti. Nunc feugiat mi a tellus consequat imperdiet. Vestibulum sapien. Proin quam. Etiam ultrices. Suspendisse in justo eu magna luctus suscipit. Sed lectus. Integer euismod lacus luctus magna. Quisque cursus, metus vitae pharetra auctor, sem massa mattis sem, at interdum magna augue eget diam. Vestibulum ante ipsum primis in faucibus orci luctus et ultrices posuere cubilia Curae; Morbi lacinia molestie dui. Praesent blandit dolor. Sed non quam.

Theory Results Lorem ipsum dolor sit amet, consectetur adipiscing elit, sed do eiusmod tempor incididunt ut labore et dolore magna aliqua. Ut enim ad minim veniam, quis nostrud exercitation ullamco laboris nisi ut aliquip ex ea commodo consequat. In vel mi sit amet augue congue elementum. Morbi in ipsum sit amet pede facilisis laoreet. Donec lacus nunc, viverra nec, blandit vel, egestas et, augue. Vestibulum tincidunt malesuada tellus. Ut ultrices ultrices enim. Curabitur sit amet mauris. Morbi in dui quis est pulvinar ullamcorper. Nulla facilisi. Integer lacinia sollicitudin massa. Cras metus. Sed aliquet risus a tortor. Vestibulum tincidunt malesuada tellus. Ut ultrices ultrices enim. Curabitur sit amet mauris. Morbi in dui quis est pulvinar ullamcorper. Nulla facilisi. Integer lacinia sollicitudin massa. Cras metus. Sed aliquet risus a tortor. Integer id quam. Morbi mi. Quisque nisl felis, venenatis tristique, dignissim in, ultrices sit amet, augue. Proin sodales libero eget ante. Nulla quam.

Experimental Results Lorem ipsum dolor sit amet, consectetur adipiscing elit, sed do eiusmod tempor incididunt ut labore et dolore magna aliqua. Ut enim ad minim veniam. Integer id quam. Morbi mi. Quisque nisl felis, venenatis tristique, dignissim in, ultrices sit amet, augue. Proin sodales libero eget ante. Nulla quam. Aenean laoreet. Vestibulum nisi lectus, commodo ac, facilisis ac, ultricies eu, pede. Ut orci risus, accumsan porttitor, cursus quis, aliquet eget, justo. Sed pretium blandit orci. Ut eu diam at pede suscipit sodales. Aenean lectus elit, fermentum non, convallis id, sagittis at, neque. Nullam mauris orci, aliquet et, iaculis et, viverra vitae, ligula. Nulla metus metus, ullamcorper vel, tincidunt sed, euismod in, nibh. Quisque volutpat condimentum velit. Class aptent taciti sociosqu ad litora torquent per conubia nostra, per inceptos himenaeos. Nam nec ante. Sed lacinia, urna non tincidunt mattis, tortor neque adipiscing diam, a cursus ipsum ante quis turpis. Nulla facilisi. Ut fringilla. Suspendisse potenti. Nunc feugiat mi a tellus consequat imperdiet. Vestibulum sapien. Proin quam.

Conclusions Fermentum odio eu feugiat pretium nibh ipsum. Aliquet risus feugiat in ante metus dictum. Scelerisque felis imperdiet proin fermentum leo vel. Vestibulum lorem sed risus ultricies. Etiam ultrices. Suspendisse in justo eu magna luctus suscipit. Sed lectus. Integer euismod lacus luctus magna. Quisque cursus, metus vitae pharetra auctor, sem massa mattis sem, at interdum magna augue eget diam. Vestibulum ante ipsum primis in faucibus orci luctus et ultrices posuere cubilia Curae; Morbi lacinia molestie dui. Praesent blandit dolor. Sed non quam. In vel mi sit amet augue congue elementum. Morbi in ipsum sit amet pede facilisis laoreet. Donec lacus nunc, viverra nec, blandit vel, egestas et, augue. Aenean quam. In scelerisque sem at dolor. Maecenas mattis. Sed convallis tristique sem. Proin ut ligula vel nunc egestas porttitor. Morbi lectus risus, iaculis vel, suscipit quis, luctus non, massa. Fusce ac turpis quis ligula lacinia aliquet. Mauris ipsum. Nulla metus metus, ullamcorper vel, tincidunt sed, euismod in, nibh. Quisque volutpat condimentum velit. Class aptent taciti sociosqu ad litora torquent per conubia nostra, per inceptos himenaeos.

References and Acknowledgements Mollis aliquam ut porttitor leo a diam sollicitudin tempor. Fermentum odio eu feugiat pretium nibh ipsum. Aliquet risus feugiat in ante metus dictum. Scelerisque felis imperdiet proin fermentum leo vel. Vestibulum lorem sed risus ultricies. Condimentum lacinia quis vel eros donec ac odio

Figure 4.3 In this poster the author has included too much text so the font size had to be reduced. Wide columns make the text harder to read and less use of colour makes the poster less attractive.

summary at the start of your poster — this is designed for the people who are interested enough to look at the poster if the title has attracted them, but who may not yet be committed to reading the whole thing. You don't have to stick to the text of the Abstract that you submitted — that serves a rather different purpose (letting the conference organisers know what you do and putting information about your work into the Abstracts book) and anyway it's likely that your work has moved on since you wrote the original Abstract.

It is a good idea to put the Abstract or summary into a bigger font than the rest of the text to indicate that this is the most important part to read. Try to aim this summary at people who are not too familiar with your research topic and see if you can convey something about the motivation for your research and the significance of your results.

As with preparing text on slides for a talk, it's really important to keep the amount of text to the minimum necessary to convey the critical information to your reader, and to convey this information in a logical order. Keep your sentences concise and to the point. Don't ramble. Keep paragraphs short. Don't have long blocks of text: they are unreadable on a poster. Use bullet points if this helps to structure the information and makes it easier to understand. Avoid acronyms if possible but if you feel you have to use them, don't forget to define them the first time they appear. Clarity should be your prime concern.

Avoid the temptation to include lots of technical details and equations on your poster. These are hard to understand and appreciate in a few minutes while standing at the poster board, especially if the poster session is later in the evening, after dinner, with drinks, ... Remember that if someone is interested they will look at your papers and all those details will be set out clearly there. On the poster you need to concentrate on the key messages only.

Even if your subject is a theoretical one, you still need to be careful how many equations you include. A poster with too many equations will look intimidating to most people and it's hard to take in the content of a complicated equation quickly. So if you do need to include equations, make sure that you stick to the most important ones only, and present them clearly (not forgetting to define the symbols used).

Figures are very useful on a poster as they convey lots of information and people will often just look at the figures without reading much of the text. A poster without any figures will look very dull. So make sure the figures are used effectively, convey a clear message and are self-explanatory. Include a caption to explain anything that is not obvious and to guide your readers to show them what they should be looking for. If you use a series of boxes for separate topics in your poster as suggested above, you can put one figure in each box and use the text in the box as an extended caption for the figure. It's generally a good idea to make the figures big as you will probably spend a lot of time pointing to details on them.

The principles that go into the preparation of the figures are the same as for figures on slides for a talk (see Chapter 6). The most important points are to make the figures simple, readable and instructive. Don't try to put too much information on a single figure — keep the density of information manageable. Use colour freely to distinguish different data sets, experiments, models, fits, etc. Pay particular attention to the size of axis labels and legends as the default sizes for these in many plotting packages are rather small.

If you take figures from your published papers, make sure they do not include extraneous labels or annotations that were fine in the original paper but make no sense on the poster — such as Figure 1A, B, C if you are only need to show A and C here, without B. And give a reference to the paper they were taken from. If you show a diagram or a figure from someone else's paper, make sure you give the complete reference to that paper in the caption.

Don't forget to include a few references at the end of your poster. This is a good way of pointing people to your latest papers. You might also cite key papers from other authors but you don't need to try to cite all relevant papers as you would in a thesis or a journal paper.

Finally, thank the organisations that have funded your work and include their logos on your poster — and thank anyone else who has helped you.

Once you have drafted your poster, show it to your co-authors to make sure that they are happy with the content. It's also worth showing it to friends and colleagues for their comments. After you have received feedback, you can make any necessary changes. Remember though that the final decision about content and style is yours, because it's you that is presenting the poster — unless of course there is any dispute about the science with your co-authors, in which case you have a problem!

4.4 Practical Issues

Poster boards have a range of sizes and shapes so check both the size and orientation required so that your poster fits the board properly. This also affects how much text you can fit on the poster of course, so you need to be aware of the poster size right from the start (Figure 4.4).

There are many places nowadays that can print posters professionally and many of them will be able to do it with very little notice, but make sure you are aware of your printer's deadline so you don't end up struggling to get the poster printed in time.

Both of us have had PhD students in our groups who have missed deadlines for printing their posters (and everything else) and have taken their posters on memory sticks to their meetings — then spent considerable time in strange cities on other continents hunting for someone to print their posters. This is not the most effective use of the opportunities offered by the conference.

Also use the poster template provided by your University or Institute if there is one. Many places have strict rules on how their students/staff can present posters. Have a look online.

Getting your poster to the conference can sometimes be a pain. You definitely need to use a proper poster tube if you are taking it on a flight, as folding it up will ruin its appearance. But watch out for airlines that charge you extra for an extra piece of hand luggage! An alternative is to use a cloth poster. These are more expensive to print but are much easier to carry and the quality can be very high. And for smaller meetings, don't forget to take suitable fixings for your poster

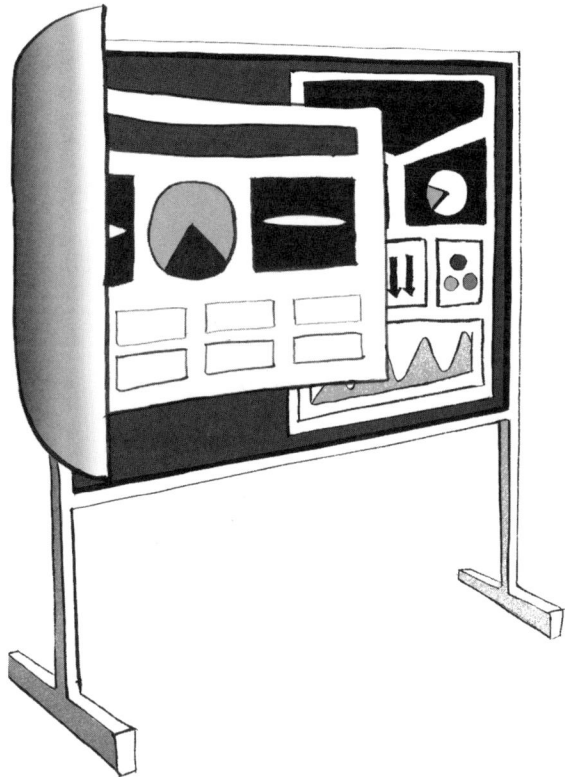

Figure 4.4 Before you make your poster, check with the conference organisers for the size and orientation required so that your poster fits the board properly.

(such as drawing pins) in case the conference organisers don't provide them.

Key Points

- Look at posters around your department to see what works well.
- Use your University/Institute poster template if there is one — check online.
- Check the right size and orientation for your poster.
- Write clearly and concisely and don't use too many words.

- Include an Abstract summarising what you have done.
- Use graphics with captions that are self-contained to get the message across.
- Use BIG fonts, i.e. at least 24 points and above throughout that can be read easily when standing a few paces away from the poster.
- Don't include lots of technical details or equations. They are in your papers and your brain.
- Columns often work well to make a poster readable, or boxes for each subtopic. Use whatever you find gets your message across well.
- Include contact details and acknowledge funders.
- Give a few key references.
- Get your poster ready in good time to allow for printing.

Chapter 5

Preparing a Formal Talk

5.1 Introduction

Most of us don't like giving talks at first, but most of us find that the more we do, the easier it gets. Eventually we reach the point where it can actually be rather enjoyable, especially if we have nice results to present to an interested audience (of course that's the ideal situation, and life isn't always like that).

In giving a talk we are communicating, conveying information to other people about what we've done or found out that will be of interest to them. We may be terrified of talking, but there are some actions we can take to give ourselves the best chance of surviving the ordeal, and the audience the best chance of understanding what we are trying to say.

The key to it all is ...

preparation, **preparation, PREPARATION**.

And this needs to be done in good time, *before* you set out for your conference, so that your preparation is not rushed. A presentation that is prepared on the plane, or, worse, the night before the talk takes place, really won't go as well as it should do. Too little time spent preparing the talk is likely to result in poor structure and unclear slides, and you will be less confident delivering it because you will not be familiar with what's in it. Sadly, at least one of us has been in this situation more than once and has bitterly regretted it.

5.2 Drafting What Goes Into the Talk and Thinking About the Number of Slides

The very first issue is: find out how long you have been allocated to talk for.

Most sessions are timed so that you talk for a certain number of minutes and there are 5–10 minutes for questions afterwards. Find out exactly how much time you have and then you can plan what you need to say and how to say it.

Then write down a list of points you need to communicate. You need to introduce your topic so that your audience understands which area you are working in and why it is important. How you do this will, of course, depends on your audience. You need to tell the audience about why and how you have undertaken a particular set of experiments or calculations; you need to talk about your results and conclusions, or, if it's early days, what you hope to achieve; and perhaps some of the possible pitfalls. Then you need to go back to the beginning and remind the audience why your work is important. Finally you need to thank, to acknowledge, all the people and funders who have contributed the work.

A rather shorter way of saying this is: **Why? How? What? Why? Who?**

- **Why** are you doing this research?
- **How** are you going about it?
- **What** have you found out?
- **Why** are your results important?
- **Who** gave you the resources to do it?

In a 40-minute talk all of this may be easy to do. In a 5-minute talk with a maximum of three slides, you may have to summarise rather a lot …

So plan the key points carefully and tell your story in a logical order, paying attention to the flow of your argument or narrative. Start by drafting your slide presentation: if you need a rough guide, the number of slides should be no more than the number of minutes you have to talk, and preferably quite a bit fewer to give time for each slide to be fully discussed and the audience to follow what you are saying. This will also depend on your personal style. As you become more experienced you will be able to judge better how many slides you are likely to need for a given length of talk.

Start with a title slide. As well as the title of your talk, this will give your name, affiliation and contact details (including your group's web page). It could also list the names of your collaborators. It is common practise to put the logos of your institute and your funders on this slide as well.

After the title slide, start if you can with a slide that sets out the structure of your talk. This can just be a list of sections or topics. It's helpful for your listeners to have some idea about how your talk is structured because it means that as they are listening they can better appreciate how it all fits together. If you are really pressed for time you may not be able to include a slide like this, but even so it is helpful to say something about the structure of your presentation (or you may be able to list the sections of your talk on the title slide).

During the talk, especially if it is a long one (say more than 30 minutes), it is helpful to refer back to the overall structure of the talk when you move from one section to the next. For example, you could have a title slide for each major section of the talk, or you could highlight where you are on the contents slide you showed at the start. Alternatively, some people just have a running header on the slides displaying the current section title. All of these things help signpost to your audience where you are in the talk and makes it easier to follow.

You will usually start with some sort of introduction. Pitch your introductory slides at the right level for your audience. Decide if you are targeting experts in your area or a fairly broad group of people. Generally it is best to err on the side of giving more introduction rather than less, if you have time. This is partly because people often don't understand as much as you may think, and partly because even the experts may have had a large lunch with slightly too much red wine, and might need a bit of help in being reminded of what you're doing. We have both sat in talks where the speaker says something like

'As I am sure you already know, ...' and we have wanted to shout out 'No I don't!'.

When you get to your methods and results, go through each in a logical, not chronological order. That is, present the facts in the way that best tells the story of your investigation, which is not necessarily the chronological order in which you did the experiments. Don't give too many technical details. These can get in the way and make it more difficult to follow the talk because it's hard to present technical details in an effective manner. It is usually more interesting for your audience to see the basic approach, the main results and the conclusions that can be drawn, rather than the technical details about how you did the research.

Include a few references to your papers, or to other papers in your field, that people can go to if they want more information. Also, if you or your group has a poster at the conference, then cite that poster for the same reason — say what the number of the poster is in the Abstract book and when it's being presented.

Don't forget to acknowledge your collaborators and funders. It's usually most appropriate to do this either at the start or at the end of your talk, depending on the conventions in your field.

If you can, use your last slide to give a short summary of the key points and to give details of any related poster or talk. Remember that this slide if likely to be visible during the questions, so put some useful material on it that can be the 'take-home message' of your talk. At the end of your presentation file, you may want to include a slide or two that specifically address questions you think the audience is likely to ask. You don't show these slides unless those questions are asked, but it may be useful to have them available.

For very short talks, where you may have as little as 5 minutes and three slides in which to present your life's work, then the introduction is on the first slide, perhaps with initial results, the key point about your results is on the second slide, and the third slide shows your conclusion and why it is important. At the bottom of the third slide you may also add names of people/funders you want to thank but you don't necessarily have to read these out.

If you are talking at a summer school or similar type of event you should, as always, make your slides clear and self-explanatory.

When preparing a talk like this, remember that you will need to leave lots of time available for questions and discussion during or after the talk. It's better to cover a smaller amount of material well than a larger amount of material in a hurry, so set yourself a realistic pace that will allow you to explain things clearly without having to rush. Try not to put too much on each slide so people can concentrate on one thing at a time. Include lots of references to places where people can find out more details. Your audience will probably be taking copies of the slides away with them after the School and using them to learn more about your topic, so it's important to make them as good as possible pedagogically for people who will be re-reading the slides without you being there to explain them.

Similarly, if you are preparing a plenary talk at a wide-ranging conference, remember that many people in your audience will not know very much about your area, so make sure you include enough introductory material that these people will be able to take something away from the talk. For a plenary talk you should really be aiming your presentation more at the informed non-specialist than the expert in your particular area. Getting this right requires careful judgement.

Presenting seminars or colloquia at other institutions also requires careful preparation, and it's worth asking the person who has invited you who will be there. A research seminar could be quite technical if the only people present are specialists in your area, but there may be other people working in related (or even unrelated) areas who will get lost unless you provide sufficient background material. There may also be research students present who would really appreciate a gentle introduction to your research. A more general departmental colloquium should be aimed at a wider audience from across your discipline.

There is always a temptation to reuse talks that you have prepared previously, to save the effort involved in writing a new talk, but this can be dangerous, especially if the new talk is for a different type of audience or is a different length. One risk is that the level of detail or assumed knowledge is wrong for your audience; another is that if you take sections out or add in new sections you may lose the original logical structure of your talk. Finally, you have to be aware of the danger of always adding in a little bit more material — perhaps new results since you gave the talk before — resulting in a talk that is too cramped and has to be rushed. So by all means reuse slides where it's appropriate but just make sure that your new talk comes over as a coherent whole.

5.3 Designing the Slides

There are a number of programs available for making presentations. Many people use PowerPoint because it's probably the program that's most widely available. However, others in the physical sciences may use LATEX. Users of Macs may use Keynote. There are other less common packages available. Whichever system you use, make sure you get familiar with it and check that it's compatible with whatever computer you will use when you are presenting. Sometimes you may be requested to turn your presentation into a PDF, which is probably the safest way to make sure that it always looks like you intended it to.

At any single moment, everyone listening to your talk is: (1) reading your slides; (2) trying to process the written text; (3) listening to what you are saying; (4) trying to process what they're hearing; (5) reading the Abstract book; and quite probably (6) trying to stay awake because they didn't finish partying until 4am/worrying about whether their grant application will be funded/wondering if they can get their children into a good school locally/considering how to sit next to that rather nice postdoc at the Gala Dinner. There is a *lot* going on at every single moment of your talk.

And that is why your slides have to have maximum impact and can only do so if they are **simple and well-designed** (Figure 5.1).

Figure 5.1   Simple slides get the message over best.

Words

Each slide has to convey a maximum of one main message, which you might want to turn into the title. Or two if you are really pushed for time. And this should be in the fewest possible words, so the audience can concentrate on what you are saying. The words on each of your slides serve two purposes:

(1) They convey the message you are talking about clearly and concisely.
(2) If your mind goes completely blank you can get clues from the slides.

If English isn't your first language, and your accent gets stronger as you get nervous, then keep your message on the slides, but in a simple way people can easily follow, **not whole sentences**, because no one can *listen* and *read* and *comprehend* all at the same time. Even if English is your first language, remember that there are likely to be non-native English speakers in your audience, and they will rely more on the text in the slides if it's hard for them to follow the spoken English.

Be careful with your use of abbreviations and acronyms. A common acronym for you may be unfamiliar for someone else, and that can render a slide completely unintelligible.

One of us attended a highly theoretical talk where the title consisted of acronyms that were unfamiliar and were never explained at any point in the talk. Not only was the title therefore unfathomable to this listener, but it set the scene for the whole talk, which brilliantly succeeded in conveying absolutely no information.

When you use bullet points (and we like bullet points) then try to have five or fewer per slide, and each bullet point should be short, just a few words. Bullet points work well because they are easier to read than continuous text in sentences.

Titles

Use the title of your slide to state what the data in the slide are illustrating. For example instead of titling a slide 'Results', and then illustrating your new finding that Landfish feathers turn from red to gold as they age, title the slide 'Landfish Feathers Turn Gold with Age'.

Figures

Use pictures and graphs to show your data, and make sure they are well-labelled to illustrate your point. That means using arrows to point out key details, and remembering to label both axes of graphs. It is easy to produce figures that are far too small to be read by the audience, especially as the default font sizes for labels and axes on most graphics packages are designed for printed figures on a page and not for a presentation. So make sure that if you stand at the back of a meeting room, you can read your own slides. You may not have any idea of the real meeting room in which you will be presenting, or the size of screen, but at least if you've checked in a meeting room in your own institute, and your figures and graphs are readable, that is a good sign.

If you are contrasting different models or datasets in a graph, ensure that the different lines or symbols are clearly distinguishable AND that there is a clear legend in the figure. We have all seen plots where it is impossible to tell what it is we are meant to be noticing.

Use colour to illustrate your points. And bear in mind that some people are red-green colour blind, so try to avoid using red and green to illustrate contrasting points. If you are presenting the same subjects under different conditions, then always use the same colours, in the same order. If appropriate present your controls first, so they give the baseline from which your experimental subjects vary.

For some figures, it can be difficult to get enough contrast to illustrate the points you wish to make (histology slides for biologists can be an example of this problem), so you may need to consider how best to present your work, perhaps by just presenting a blow up of a small area.

Often it's a good idea to try to include some sort of figure on every slide if possible. This gives a bit of relief from text and can make the slides look more interesting. But don't fall into the trap of putting something in for the sake of it: only include a figure if it adds something to your message. Irrelevant pictures are distracting for your audience.

Chapter 6 discusses the preparation of figures for talks and posters. There are many books and articles that give more advice about how to prepare effective and scientifically sound figures. Our own book, *Enjoy Writing Your Science Thesis or Dissertation!* (Imperial College Press/World Scientific, 2014) has a long chapter devoted to this topic. The main points that we discuss in that book are as follows:

- Label each axis clearly.
- Use SI units wherever possible.
- Use clear symbols for all the data points.
- Include a line showing a fit to the data if it's appropriate.
- Don't include wavy lines joining up the data points.
- Include error bars if necessary.
- Use colour to distinguish between different datasets.
- Don't include any spurious information on the figure.
- Overall, make sure your graphs and figures look professional.

Animations and videos

Do not put in extra animations if you don't need them. We have seen presentations in which data come sweeping in from the left, falling down from the top, rushing in from the right, leaping up from the bottom, or popping into the centre of a slide. This has two effects: (1) the audience focuses on the animation, not the information it is trying to convey; (2) several members of the audience will leave because they feel sea-sick. There are some presentation packages where the presentation consists of zooming into elements of a main slide to show more and more detail. Some people find these presentations very disorienting and completely distracting (including at least one of us).

One useful way to use animation in a presentation is to have successive bullet points (and related images) appearing one at a time. This helps the audience to concentrate on what you are saying at the time. Alternatively, you can set your presentation to have all the text there the whole time but to highlight the part you are talking about at the moment. Another helpful use of animations is to build up a diagram or figure one step at a time. But the use of unnecessary animations can distract your audience from what you are trying to say, so it's best avoided.

Videos in talks can look wonderful. They can be exciting and the perfect illustration of what you want to say. Everyone loves videos. But use them sparingly because they should only be in your talk if there is a reason to include them. Not just because they look pretty cool …

If you do have animations or videos in your slides, check that they work on the computer you will be using at the meeting, *before* you come to give your talk. This is an important point. Even if you are using the same program, the details of how your presentation comes out can change when you change computer, especially if the font or media player you are using is not installed.

One of us gave an important presentation to 200 undergraduates which hadn't been checked on the lecture theatre computer. Unfortunately all the bullet points came out as little steam trains, which didn't encourage the students to take the presentation very seriously.

Backgrounds

Most universities or research institutes have a template that you are supposed to use as the background to your slides. This template usually just includes the name or the logo of the university/institute. This is a means of publicising where you have come from and it reinforces the 'brand' of the institute.

If you do not use a template, use one plain single colour only in the background. If you have wavy lines, stripes, sunsets, background pictures, etc., you immediately start to lose the audience while they ponder if these backgrounds have any significance for the talk and then wonder about your colour sense. No busy backgrounds please: keep the background to simply one colour only. Page numbers can be useful — if only so you have an idea how far you are into your talk!

Colour and fonts

The colour scheme you use is generally up to you, except in the situation where you are presenting as one of a group of people from one institute/department, in which case you may be asked to format your slides in the same way as everyone else, to show continuity and uniformity.

If you have a free hand, then generally black text on a white background is simplest and most readable, or yellow or white text on a black or dark blue background shows up most clearly.

You can use fonts of different colour and size to illustrate different levels of information. For example, if you are working on a black background, your title line might be yellow, whereas points about data may be in white. Different levels of bullet points can be put in different colours. Try not to use colours that are too garish as that can be harsh on the eye (Figure 5.2).

Fonts: what not to use and what to use. Many people recommend using a sans serif font like **Arial** or **Calibri**. **Times New Roman** is also one of the most common fonts; **Comic Sans** is friendlier and less formal but some people don't like it; 𝔒𝔩𝔡 𝔈𝔫𝔤𝔩𝔦𝔰𝔥 𝔗𝔢𝔵𝔱 is not generally a good idea. Look at what other people use and make your own decisions about what is clearest and easiest to read. One of us tends to use **Arial** or **Calibri** most of the time.

It's important not to use a font size that's too small or too big. Something like 18–24 point is probably about right, depending on the style of the font and the screen size. This works out at a *maximum* of about 10 lines of text on a slide. Fewer lines are generally better. Remember there is no point in including a lot of text on a slide, whether it is in a sentence or on a figure, if no one can read it. That just irritates the audience.

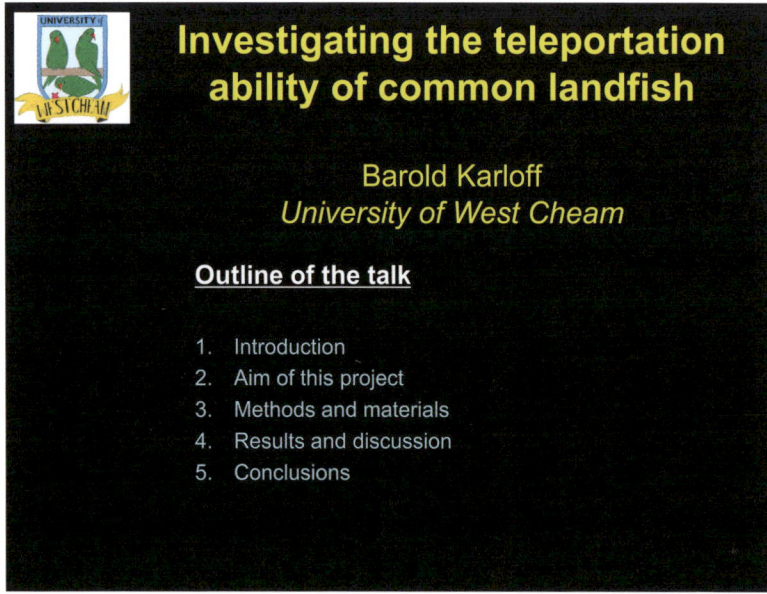

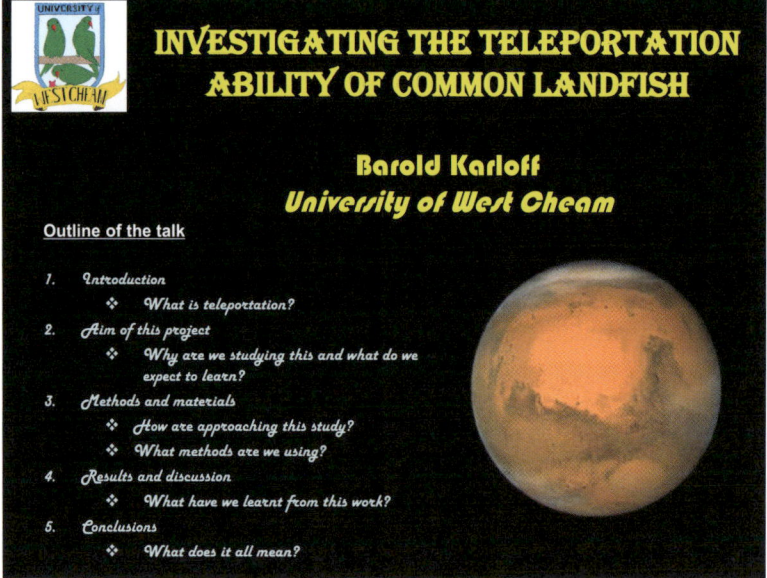

Figure 5.2 An example of a good and a bad slide giving the outline of a talk. Note the use of several fonts, an irrelevant picture of Mars and repetitive and wordy bullet points about each section of the talk, necessitating a font size which is far too small.

In the old days of glass or plastic slides, one of us was a meeting in New England, USA where a distinguished British biologist wanted to talk about some new data but hadn't had time to make their slides. The distinguished biologist gave a talk in which the data were projected in enormous metre high letters that looked like a font no one had ever seen before or would ever use again — because the distinguished biologist had taken a felt tip pen and written by hand directly onto the glass slides. Fortunately, it's not possible to do this sort of thing these days ...

As a general point, when you watch other people's talks, look carefully at how they format their slides and what you find most helpful as a listener. How much text they include on each one? Are the figures effective? Do the slides look too busy or cluttered? We can learn a lot by looking at what other people do well or badly and making appropriate changes to our own style.

Equations

It is a good principle to keep the number of equations to a minimum. It is sometimes said that you lose an audience as soon as you show an equation in a talk. While that is not going to be generally true for scientific talks, it is still the case that unfamiliar complex equations can be hard to follow when you only see them for a very short time. Equations are of course part of the language of science and so they are often necessary — indeed sometimes a simple equation can convey a concept more efficiently than words for a technical audience.

However, it is very easy to overestimate the ability of even a scientifically literate audience to grasp the significance of equations in a talk, so be careful not to include too many on your slides. If you do need to use equations, try to follow these recommendations:

- Use the minimum number of equations necessary.
- Use a large and clear font so that subscripts and superscripts can be seen clearly.
- Ensure you have a professional display equation layout by using a package such as LATEX or Equation Editor within Microsoft Office.

Don't use normal characters to create a complicated equation in a line as this will always be harder to read.

- Define all the symbols you use on the same slide as the equations.
- Explain carefully what the equations mean when you are talking.

As an example, we can show the solution of a quadratic equation $ax^2 + bx + c = 0$ in two different ways:

$$x = [-b \pm \surd(b^2 - 4ac)]/2a$$

$$x = \frac{-b \pm \sqrt{b^2 - 4ac}}{2a}$$

The first one, using standard symbols, is much less readable than the second, which uses a proper equation editing package.

5.4 Final Steps

When you have drafted your talk, and you believe you have introduced your work, made all the points that you want to, reiterated why your work is important and drafted the final acknowledgement slide, all beautifully and clearly presented with the fewest words, and simplest presentations, the nicest colours, go away and ignore it all at least overnight. Then come back to it, and either alone, or with a friend in your area, stand up and speak out loud, running through each slide, using the words you want to use in the real talk. Think about whether it makes sense to someone who knows your field but not your work. Does every slide follow on logically from the previous slide? Is it clear why you have taken each step? If you mention highly specialist items such as individual pieces of DNA or advanced fabrication techniques particular to your research field, or specialised computational packages, is it clear to the audience what you mean and why you are using these items? This is where a friend or a more experienced person such as a postdoc can be most helpful in pointing out where more explanation might be needed, or perhaps less detail may be useful (not all detail is useful or needed).

As important: are your slides consistent throughout? Same font, same background, same colours for each type of sample, same style? If not, again the unexpected differences will distract the audience from the message you are trying to tell them.

Then edit your talk taking on board the comments you have received. Be prepared to be ruthless if necessary — if it needs changing, change it! It's easy to stick with what you've got — after all, you've put a lot of effort into it — but you should make yourself rewrite parts of the talk if you need to.

Now you are ready to ... rehearse! This needs to be done *before* you go to the conference, because otherwise you will miss out on all the other things going on while you are at the conference.

And a key point here is to rehearse *and* all the time check you are keeping to time. You will get faster as you get more fluent, but always keep your eye on the clock.

Inaugural lectures for new professors have to be delivered to an audience ranging from friends and family right up to senior professors and visiting scholars. When one of us was preparing our inaugural lecture we were very worried about how it would go down. After a disastrous run-through with the PhD students in the group, the feedback received transformed the talk and it ended up much better as a result.

Key Points

- Find out how long you have to talk.
- Draft the **Why? How? What? Why? Who?** of your talk as a series of bullet points first.
- Give an introduction to your talk for people who are not specialists in your field.
- Tell your story in a logical, not chronological, order.
- Keep the slides as simple as possible.
- Use the titles to show the key point of each slide.
- Use colour to emphasise the structure of your slides.
- Make sure that the figures are appropriate and add to your message.

- Use animations and videos sparingly.
- Keep to one plain simple background.
- Check that all the slides come out correctly using the conference's computer and AV system before your talk.
- Check your font and figures are all readable from the back of the room.
- Finally, check again that you have a logical, step by step story to tell when you talk through your slides with a friend or your supervisor/line manager.

Chapter 6

Figures and Tables

6.1 Introduction

As with all types of scientific presentation, figures and tables are extremely important because that is where you present the bulk of your detailed results.

When writing a research article or a thesis, it's OK to present a lot of information in a figure or a table (within reason!) because your reader has time to look carefully at the data if they want to, though of course it's always necessary to present the important information with the minimum of clutter. Your reader can concentrate on your results, pick out small details on pictures, compare different curves on the same plot, compare columns of a table and read the captions carefully.

However, when you are writing a poster or preparing a talk, you need to be much more careful about how much detail you give because (1) your reader will not be able to see your results as clearly as they can when reading a paper and (2) they will not have enough time to study it in detail.

So the main message about presenting figures and tables is: KEEP IT CLEAR AND SIMPLE. Don't be tempted to assume that you can automatically reuse figures from your papers. These may well be too detailed for a poster or a talk. You need to look at each one carefully to assess whether it's OK as it stands or whether it needs changing. It's attractive to take the easy option and use figures you already have available, but don't do this unless you are sure that they are suitable.

We will look in turn at the main types of figures: photographs, diagrams or line drawings, graphs, and other plots; finally we will consider how some of the same ideas apply to tables.

6.2 Photographs

Everyone loves photos. A well-chosen photo can make a poster presentation look very attractive and it can make it stand out among others that are covered with large blocks of text or equations. Similarly in a presentation, photographs can make slides look much more attractive. However, they are only effective if they are (1) relevant to what you are trying to say and (2) carefully chosen to illustrate what you want to get across.

Relevance is important not just because otherwise your photograph is taking up valuable space on your poster or slide: if it's not relevant it will positively get in the way of your message because people will be looking at it and wondering why it's there, instead of listening to you. So don't put a photograph in just because it's attractive: make sure it's going to add something to what you are trying to say.

Once you have decided that a photograph is necessary, you need to make sure that it's the right one. Ask yourself what point you want to make with the photograph. Is it just to give an idea of what a piece of apparatus looks like? Is it to show a sample or specimen that you have worked with? Is it showing the results of some experiment? Do you expect someone just to glance at the photograph quickly or do they need to study it in detail?

Once you have decided why it's there, you can then work out how to use it as effectively as possible (see Figure 6.1). Here are some general points to consider:

- Keep the photograph as clear as possible with the minimum amount of clutter. Check that it's in focus and that there is good contrast. Remember that if it's to be printed the colour contrast may be different (and not so good) as on a computer screen. More importantly, the contrast on an image is likely to be significantly worse when projected than it is on a computer screen, particularly if the room is not very dark. If in doubt, check with a real projector: it can be very disappointing to see how poor a photograph can look when projected, and you don't want it to be so bad that the photograph doesn't then show what you need it to show. No-one wants to have to say 'You can't really see this on the screen but what should be noticeable is ...' in their talk.

- Choose an appropriate size for the photograph — big enough to show the necessary detail but not so big that it takes up valuable space that you could use for other things.
- You may need to label parts of your photograph to draw attention to them. Make sure that your labels are clear and big enough to be read from a distance. Use the minimum number of words necessary. Don't have too many labels because it takes time to absorb all that information.
- If you are comparing a number of photographs, try to arrange them to make the comparison as easy as possible. For instance, make sure they are all constructed the same way and that the images are the same size. Put them next to each other if possible.
- You may well need to include a scale bar to give an idea of the size of the subject of your photograph. Make sure this is clearly visible.

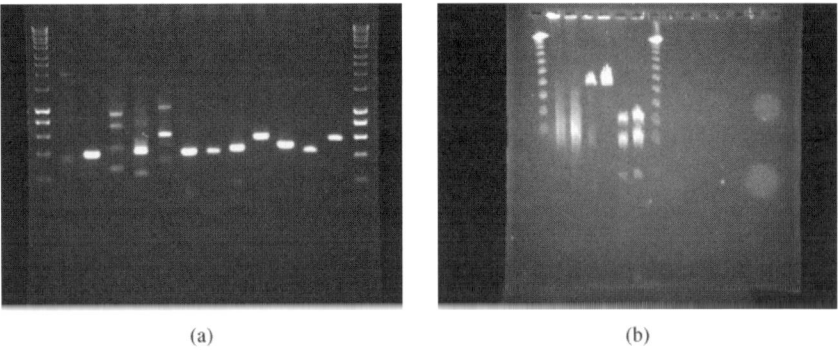

(a)   (b)

Figure 6.1 A photograph of gel electrophoresis of Landfish DNA fragments, ready to be annotated and put into a diagram. (a) the good — all data are clearly presented and beautiful; (b) the bad and the ugly — smeary data, in one part of the photograph, with two strange blobs on the right-hand side.

- Try to choose representative and nice examples of your data, so that people can see the quality of your work, without worrying whether the results are reliable because the data are sloppy.

6.3 Diagrams

Diagrams can be useful for lots of things: they can save a lot of words of description and it's much easier to take in visual information than lots of text. Some examples of effective use of diagrams include the configuration of a complex optical setup; the steps in the synthesis of a chemical; a flow diagram for a computer algorithm or a process for treating samples; the construction of an electrical circuit; and cutaway drawings of an object or device.

Remember that whether the diagram appears on a slide or in a poster, your reader will have very little time to look at it and so you have to help them to understand the purpose of the diagram as easily as possible.

As with other visual material, strip the diagram down to the basics with the least possible distractions, for example, unimportant elements or unnecessary details. In labelling parts, keep the labels short, clear and readable from a distance. One option that is often used in journal papers is just to label the diagram with letters and have a key in the caption. Where possible we recommend that you avoid this, especially for talks, and instead use words in your labels. That makes it easier to follow because you don't have to keep looking at the caption. If you need to, you can always add more description in the caption on a poster, but the basic principle should be that the diagram stands on its own without the need to refer to the text.

See Figure 6.2(a) for an example of a diagram that has a lot of unnecessary detail and Figure 6.2(b) for the same diagram with the unnecessary aspects removed. Notice how much easier it is to take in the key information from Figure 6.2(b) than it is from Figure 6.2(a), because of the clutter in Figure 6.2(a).

6.4 Graphs

The principles to follow when preparing graphs for a poster or for a talk are much the same as for papers or theses:

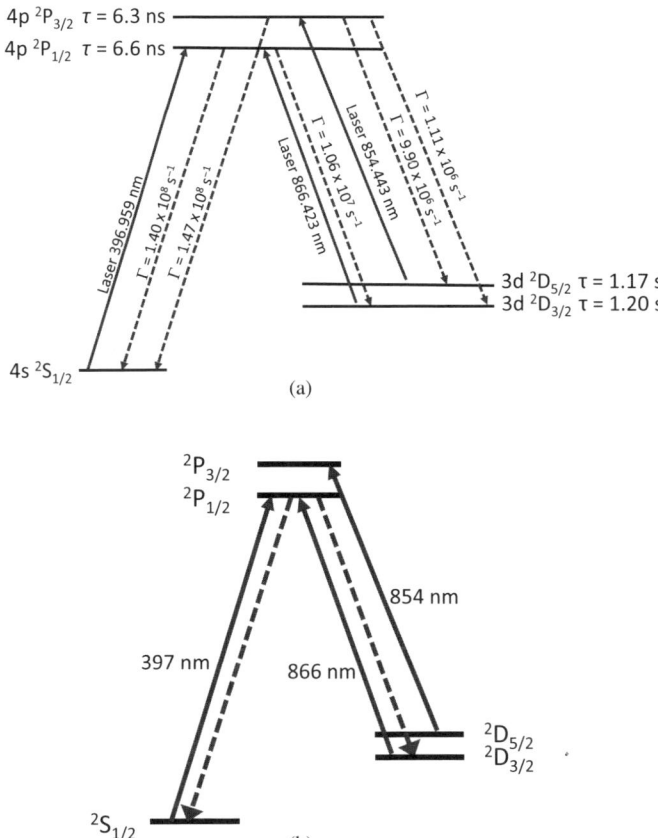

Figure 6.2 A simplified energy level diagram for singly ionised calcium (Ca⁺). The idea is to show the relevant energy levels and transitions that can occur when the ion is illuminated with lasers at the wavelengths shown. (a) Shows too much detail so that it is difficult to see the important information at a glance (e.g. the laser wavelengths are given to too many significant figures to be useful and the various decay rates given are not necessary). (b) Shows the most important information more clearly (larger font, fewer significant figures, bolder lines). The version shown in (b) is much more suitable for a talk or a poster than the version shown in (a) — which might be used in a paper.

- Don't try to put too much information on one graph.
- Label the axes of the graph carefully, using a clear font that is big enough to read easily.
- Don't forget to include units.

- Put error bars on your data if appropriate.
- If there are multiple curves, label each one clearly.

However, extra considerations come into play when your graph is for a poster or for a presentation. As with the other topics discussed above, the main reason for this is the need to make it quick and easy for your reader to understand the purpose of the graph and to take away the right message from it.

In a printed publication it may be appropriate to put several plots on the same graph. You may need to consider whether it's a good idea to use colour or not because the printed version of a journal may not have colour and even when printing from a PDF, some people will use a monochrome printer. So it pays to make sure that your plots are distinguishable even in monochrome (see Figure 6.3).

In a presentation or on a poster this isn't a concern because you know that it will be presented in colour, so it's a good idea to use colour to help make graphs as clear as possible (see Figure 6.4).

However, a plot that has so much data on it is probably going to be too much to take in, so it makes more sense to cut the amount of information down to make it more manageable (see Figure 6.5).

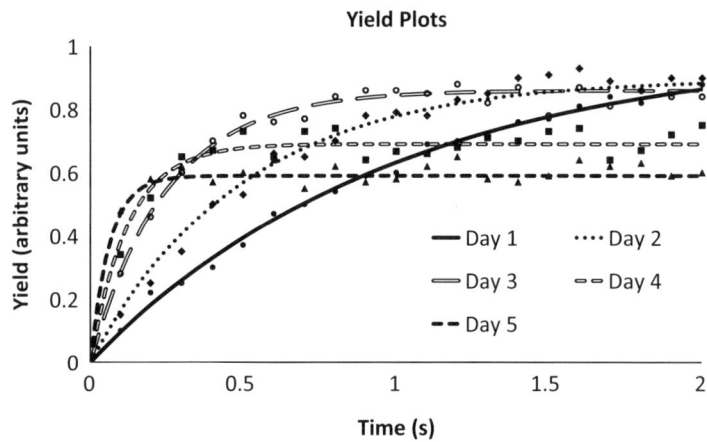

Figure 6.3   A plot that is suitable for reproduction in monochrome.

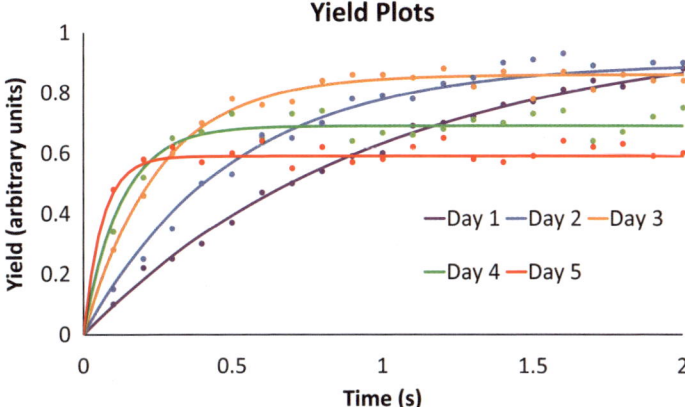

Figure 6.4   The same data as Figure 6.3 but using colour makes it more suitable for use in a poster or a talk.

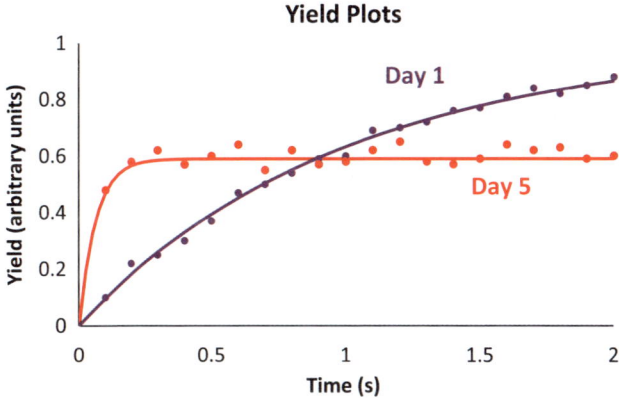

Figure 6.5   The same data as Figure 6.3 but cut down to make a single point in a way that can be grasped quickly, for example on a slide in a talk.

This still makes the main point but without all the detailed evidence. Remember that if someone is inspired by your poster or talk to look at your work in more detail they will then look at your publications and see the full dataset and complete details. So if you can't include everything in your poster or talk (and this will nearly always be the case!) don't worry. It's more important to use the conference presentation to get people interested in the first place.

Another consideration here is to make your graph tell its story as much as possible without the need to refer to the caption. This probably means that it needs more labels than a graph in a paper would typically have. Also it's not normally appropriate to include titles on graphs in papers but in a poster or a talk that can often be a good idea, especially in the absence of a more detailed caption. That is why we included 'Yield Plots' as a title in Figures 6.3–6.5, which are intended for a poster or presentation, but we would not do that for a printed article.

6.5 Other Plots

It's good to use a variety of plots in posters and talks, so think carefully about what might work well for you. For example, a pie chart might be good to show data on types of research, but the style used in Figure 6.6 makes it hard to understand the figure quickly. Figure 6.7 is much more effective because the information you want (the labels of the different categories) is right next to where it is

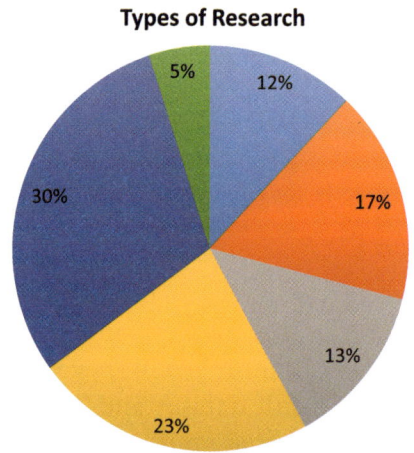

Figure 6.6   A simple pie chart.

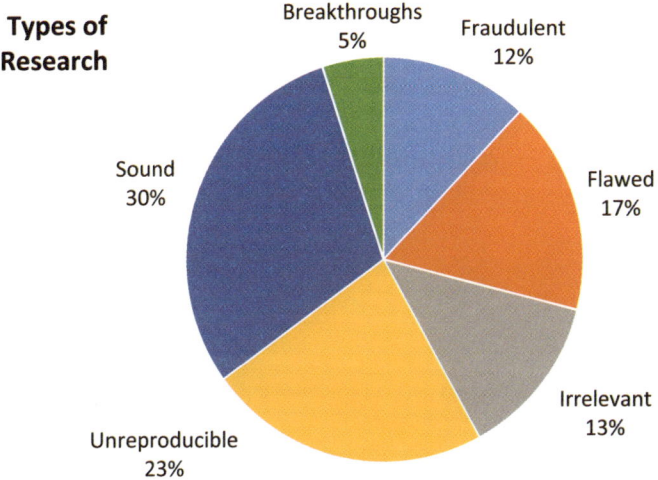

Figure 6.7   The same data as in Figure 6.6 but presented in a manner that is easier to follow.

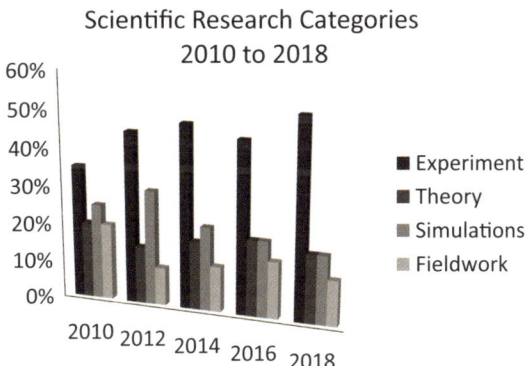

Figure 6.8   A histogram showing the changes in types of research over time. The similar shades of grey and histogram style do not make it easy to see the trends.

charted. Note also how a more appropriate font size makes this plot easier to read than in Figure 6.6.

In order to show how something is changing with time, for example the distribution of categories of research, it's probably most appropriate to use a histogram of some sort but make sure you give some thought to how this is presented. Figure 6.8 shows a mono-chrome histogram with an unnecessary 3D effect.

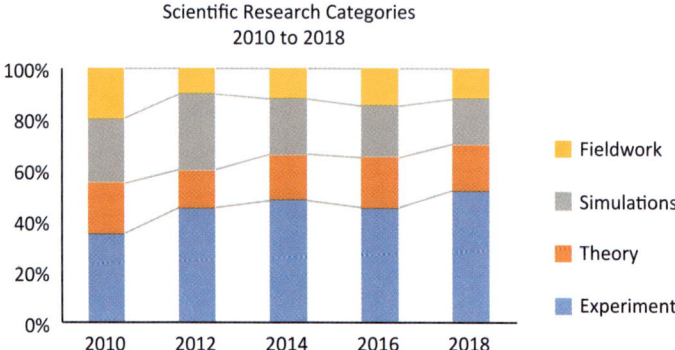

Figure 6.9   A colour stacked histogram showing the changes in types of research over time. This type of histogram is more effective than Figure 6.8.

A stacked histogram is more effective for showing the trends. Again, colour can be used to make the histograms as clear as possible (see Figure 6.9).

6.6 Tables

Sometimes it is appropriate to use tables for numerical data in talks but most of the time it is not a good idea. This is because it takes time to look at the table and scan it to see how different entries relate to each other. The example shown in Table 6.1 lists the data used for Figure 6.8 but it is much easier to understand what the data show by looking at the histogram rather than by reading the numbers in the table. However, if it is for a poster rather than for a talk, it may be more appropriate to use a table because it contains more detailed information. Visitors to the poster may have more time to read the table carefully if the details are important. But if it's just the general trend that's important, it's probably still best to use a histogram.

Tables that contain textual entries rather than numerical data are better to use in posters than in talks, in general. The amount of information you can fit onto one slide in a talk is quite limited, so that restricts the size and detail that you can present in a table. However, on a poster it's OK to have a larger table containing a lot of

Table 6.1 A table showing the data used for Figure 6.8. Note that it is much easier to gain an overview of the data from the histogram rather than from the table.

	2010	2012	2014	2016	2018
Experiment	35	45	48	45	52
Theory	20	15	18	20	18
Simulations	25	30	22	20	18
Fieldwork	20	10	12	15	12
Total	100	100	100	100	100

information if that's the best way to present it. But remember it's still hard to take in a lot of information at once so make sure that you don't include anything that's not necessary.

If you do use tables in a talk, there are some guidelines that it is best to follow:

- Use the minimum number of columns and rows as possible.
- Have clear (and short) headings.
- Don't be tempted to use a small font to fit in lots of information.
- Keep the entries short.
- For numerical entries, don't give too many significant figures.
- If there are entries that you particularly want to draw attention to, consider highlighting them.

Key Points

- Keep it simple.
- Don't reuse figures from papers if they are too complicated.
- Use colour to enhance your figures.
- Make your figures visually attractive.
- Use a suitable large and clear font for labels.
- Keep tables as short as possible.

Chapter 7

Presenting a Poster

7.1 Before the Poster Session

Presenting a poster at a conference may be your first opportunity to talk to new people in public about your research. It's quite possible that you will feel very nervous and rather inadequate for this task. Many people do. But you should remember that you are the expert in this topic and most people just want to give you the opportunity to tell them about what you are doing and why it's interesting and important. So try to relax and enjoy this opportunity!

You may find yourself talking to some of the big players in your field. This can be pretty intimidating especially if you have not had a chance to speak to them before or if you have only done so in the company of more senior people like your supervisor.

So as with all other ways in which you present your research to other people, the most important thing to do beforehand is to make sure that you are properly prepared for conversations you will have at your poster. This means thinking about what you will say and making sure that the poster has the right information on it (and that you know your way around the poster!). You also need to check that you have paper and pens so you can make notes from conversations or give contact details to your visitors.

On a more practical level, check what you will need for fixing your poster to the poster board. Normally conference organisers will make sure that they have all the necessary materials (drawing pins, Velcro strips, tape, clips, etc.) available, but it's worth having your own supply in case they don't or in case they run out.

7.2 At the Poster Session

You will have different types of visitors to your poster and we will consider them in turn.

First of all, many people who don't know anything about the subject of your research may see your poster and stop to look at it out of curiosity. In particular, at a conference with a wide subject coverage you may get quite a lot of people who are attracted by something in the title, or perhaps a figure caught their eye. They may well just ask you 'Please could you tell me what your research is about?' It's a good idea to think in advance about what you are going to say to people like this to give a summary of your poster. This is like an 'elevator pitch' — a 30-second statement that gets across the key points.

Your elevator pitch should say who you are, what research you are doing, why it's interesting and what your main conclusions (so far) are. For someone from a different field, you can only expect to convey a relatively small amount of information in this time. Use the key diagrams or pictures on your poster to emphasise the points you want to make. The aim is (1) to give this basic information to someone who is going to move on straight away so they at least think 'That's interesting stuff'; and (2) to pique the interest of people who *ought* to be interested in knowing more about your work.

Every now and then you may get someone coming along who is not really interested in your work but just wants to ask difficult questions to show off. Thankfully, this is very rare. Don't be upset if this happens: it's not your fault. Just try to finish the conversation politely so you can talk to different people.

You will also get visitors who already know about your research area and perhaps already know what you are doing, and they are likely to come along with specific questions. With these people, of course, you can probably safely jump straight into a detailed conversation because they already know all the background to your work. However, even with this type of visitor, remember that it is very easy to overestimate how much of the detail people understand, so watch out for signs that you are going too fast.

It may be hard sometimes to tell how familiar someone is with your field, in which case it's perfectly fine to ask them if they are familiar with the topic. It's much better to find out what their level of knowledge is before you launch into your explanations, particularly if you pitch it too high, because some people may feel uncomfortable asking you to go back to the start so they can ask some simple questions. Think about the reverse situation — as a visitor to a poster you probably appreciate it when the presenter makes it clear that it's OK to ask them basic questions.

Sometimes it can be a little frustrating when you spend time at your poster without any visitors, and it can leave all of us feeling embarrassed and rejected. Try not to let this get to you. The number of people who talk to you is always likely to fluctuate wildly, and you will most likely have good conversations at some point. It may be just because your particular research turned out to be a little bit unusual for the conference you chose, and in that case you may need to choose a different conference next time. But don't worry, because you will still benefit from attending this one for lots of other reasons (Figure 7.1).

It can be even more frustrating if you don't have any visitors for a long time and then suddenly you have too many all at the same time. You don't want anyone to walk away because there are too many people already talking to you, but if they do walk away it's quite likely they will make a point of coming back later to see if you are not so busy.

Finally, there will be occasional people who come by your poster, because they are at the meeting on their own, and are shy, and you are there, and look nice, and they can ask you a question and you will be pleasant to them, and they can stop drifting around the poster session feeling socially isolated. This is good for you, as you have an interested and willing audience, and often encounters like this can lead to life-long friendships. So be generous with your attention. Equally, if you're feeling socially isolated because no one has come to your poster, don't desperately hang on to the one person who swings by and hold onto them conversationally for the entire evening. Allow people to flow around you.

Figure 7.1   Early in the poster session, Agnieszka thought that no one was interested in her work and felt a bit miserable, but these things are always unpredictable … and shortly afterwards she had lots of people wanting to know more.

Of course if you are standing at your own poster it means that you are not able to look at the other posters, and it's very likely that the other posters being presented at your session are ones that you particularly want to see! If you have a colleague with you who is a co-author on your poster, you can arrange to split the time between the two of you so that both of you get a chance to look at the other posters. You don't both need to be present at the poster the whole time.

Even if you don't have a co-author present but there is someone else from the same research group, they may be prepared to 'hold the fort' for you for a while so you can look at other posters. If someone visits who wants to hear more detail, your colleague can tell them when you will be back so you can have a conversation with them later.

If you don't have anyone who can cover for you, then you will have to make a decision about whether you can leave the poster unattended for some time while you walk around the rest of the session. It's probably a good idea to do this as posters give an excellent opportunity to go and ask other people about their work. But in order to use that time most effectively, it's best to plan carefully which posters you want to go to in advance. You could even leave a note to say exactly what time you'll be back and available — but then make sure you really are back by that time.

A final comment about poster sessions: they are often accompanied by drinks and perhaps nibbles, especially if they are held in the evening. That's fine and it's nice to have a drink while talking about science, but be careful not to indulge *too* much in the free alcohol! If you are meeting potential collaborators or competitors (or even employers) you definitely want to be on your best form and not to have your senses dulled by having too much to drink. Keep it professional: save all that for *after* the poster session.

7.3 Dealing with Questions

If you have successfully gained the interest of your visitor, they will then start to ask questions. This is your opportunity to really engage with someone about what you have been doing. It's a good idea to try to pick up clues to work out how detailed their knowledge already is about your topic. This will help you to answer questions on the right level of detail. But in general, if you're not sure how much they know, it's best to assume that the questioner doesn't know very much and to give straightforward answers that start from relatively basic knowledge. Don't be surprised if you need to repeat some of what you have already said — it can be very hard to take in a lot of new ideas in a short time. At this stage, don't go into too much detail. If people want to know more, they will always say so. But if you jump in with detailed answers which your visitor doesn't understand, they are likely to be put off and they may well move on because they are embarrassed about asking you for more basic explanations. Keep eye contact with your visitor and check their body language as you are explaining things to them. If you can see that they are beginning to glaze over, adjust your explanations accordingly.

If one of the 'big cheeses' for your subject area comes along, it can be stressful but you should remember that they, like everyone else, just want to find out what you are doing. When they ask questions, they are genuinely interested in the answer and they are hoping to learn something new from you. But even with these people it's possible to overestimate what they already know. They may be very knowledgeable about the field as a whole, but they are not an expert in *your* work, and you are! So you can have very good two-way conversations with experts that are beneficial for both of you.

It can be really stimulating to get into a detailed conversation at your poster. This, after all, is one of the main reasons you came to the conference! It's a chance to tell someone about what excites you scientifically and gets you out of bed in the morning. Many of these conversations don't lead to any further contact but some may lead to new ideas for you to try out, or future collaborations, or even a new job when the time comes to move on.

If you are fairly new to your project, you may well find that there are some questions to which you don't know the answer. This is perfectly normal, and you shouldn't worry if this happens to you. It can happen to anybody, whether an established researcher or new to the field. Because they are seeing your work for the first time, your visitors may well come up with questions that you have not thought of before. You can always take their contact details and get in touch with them after the conference when you have had a chance to think about their question some more, or when you have been able to discuss it with others. This is one of the benefits of going to a conference: to help you to think in new ways about your project and to get new ideas, which can come from you or from other people.

7.4 After the Poster Session

After the poster session, you'll probably be exhausted! And that's a very good sign because it shows that you have been busy. Now you can afford to relax a bit more.

It's worth taking notes during the poster session if there is anything that you need to do later, for example, to follow-up on a conversation or send a paper to someone who asked a question or look at papers from a research group you weren't familiar with. Don't rely on remembering things like that — if you are like us you will definitely forget if you don't write yourself a note.

Key Points

- Prepare a 30-second elevator pitch before the poster session.
- Don't assume that your visitors already know a lot about what you are doing.
- If in doubt, start your explanations at a basic level.
- Look for signs that show if your visitors are understanding what you are saying.
- Enjoy the opportunity to talk about your work to an interested and well-informed audience.
- Follow-up on conversations after the poster session.
- Bring something to note down your name/email address, and give it to people, and vice versa, for following up on interesting points — or a new social life.

Chapter 8

Giving a Talk

8.1 Practise, Practise, Practise!

The most important point about giving a talk, especially if you are nervous or you haven't given many talks before is to practise, **Practise**, **PRACTISE**.

And to keep practising until you are completely fluent. We do. We still do. Both of us. And we've been doing this for decades ... and by 'practise' we mean **SAY THE WORDS OUT LOUD**. Saying the talk out loud, even to an imaginary audience, is incredibly important so that you start to automatically remember the words and get a feel for timing and you start to feel confident with what you're saying, or so you can change what you're saying if you realise it doesn't work.

If you are fluent you can be confident you are talking to time, and even if you completely forget absolutely everything you have to say, you know that the clues from your slides will get you through.

If you are nervous, practise especially your first two sentences. Say them over and over and over until you are completely bored with them. Even if your first sentence is 'Hello, my name is ...' So when you get on stage you will be on automatic pilot and can simply start speaking.

Practise at first on your own, standing up, with the real slides being projected by a real slide projector and, again, talking out loud. We do this. People walk by the room and see us through the glass panels in the door and instead of thinking *Professor Thompson or Professor Fisher are completely nuts, they're talking to an imaginary audience,* they think, *Oh, someone practising for a talk.* Trust us on this.

It's important to say the words out loud, at the pace you will be delivering the talk. Don't rush. Then when you say the words you get used to what you're saying, and perhaps can even start to streamline the talk if necessary.

When you have practised your talk a few times ask a friend, or a couple of friends, to listen to it. Ask them to check they can understand what you want to say, and that your slides make sense. Act on what they tell you. Everyone will have a different point of view and you can work out what you think is important and which suggested changes you need to undertake.

Every time you practise, you must stick to the right time for your talk, so set a timer and make sure you get used to the timing. If you find you are going over your allotted time, *cut stuff out!* Don't just try to speak faster.

Finally, load your talk onto your computer or USB stick and take it to the meeting. Just in case, email it to yourself so that you can access a copy if anything goes wrong. If possible also email it to the organisers so that they can load it before you arrive, if they are willing to do this.

8.2 Scoping Out the Audio-visual (AV) System

The day arrives. You are talking today. You have practised your talk. What do you need to do next? Start by getting your presentation loaded onto the meeting computer as soon as you can. Or if you are talking from your own computer, make sure everything is working properly. Check that all the slides show up and that the animations are working. Do this with enough time so that you can correct any problems. For example, if you are talking in the morning, do all this the day before if you can.

It is important to check that your talk works properly on a different machine than you use at home if the organisers provide the computer for presentations. If you are using your own laptop, remember that you may need adaptors to connect it to the AV system (e.g. if it is a Mac), so don't forget to take them with you! Also if you are using your own computer make sure it is fully charged or that you have a power lead with you.

If you can, bring a laser pointer with you. Most major meetings in large well-staffed venues will have a laser pointer separately or integrated into the AV equipment. Many smaller venues will not, and laser pointers are useful for drawing attention to important facets of your data. Even better, use a 'presenter', which is a laser pointer that can also control your presentation remotely. Then you don't have to stay near the computer but can advance the slides from wherever you are. This can really help you to give a smooth and professional presentation, so long as you don't fiddle with the presenter and accidentally advance the slides at the wrong time!

After you have checked that your talk has loaded properly then scope out the AV. Make sure you know how to forward and reverse the slides, and how the laser pointer works (if there is one), and how to use the radio microphone if necessary. Use the time in the breaks before your talk to look at it all a final time, and see what the other speakers do when they talk. This is taking a professional approach to talking, and this preparation is important to avoid embarrassing delays when you go up to speak.

Be prepared; one of us bumped into a friend, a well-known biologist, who was talking at the same session. The biologist was an honoured speaker, and due to give a 40-minute talk and this person had brought along entirely the wrong slides.

8.3 What to Wear

It seems like a trivial point, but looking and feeling good will help you to be confident when you are giving your talk. It's not a fashion show, however, it's worth thinking in advance about what you are going to wear.

For most scientific conferences the dress code is generally casual. It's best to wear fairly neutral and comfortable clothes that you can forget about and don't need to continually adjust. Avoid potential wardrobe malfunctions like bursting buttons, trousers that are too loose, etc. And don't fiddle with your clothes (or hair), as this can be very distracting for your audience.

Remember that when you are talking, you are quite likely to get hot and sweaty, particularly if you are very nervous. So take off your jumper or jacket if there is a risk that you will otherwise get too hot.

Of course there are some conferences where the dress code is much more formal, for example, official state-sponsored events, so do your homework to make sure that you don't misjudge the appropriate clothing to wear.

Finally, you may want to bring a bottle of water with you onto the podium, so you can drink it if you need to — not all meetings will provide you with water, so be prepared!

8.4 Giving Your Talk at the Meeting

Waiting to talk can be nerve-wracking, and you may not be able to focus on what the previous speaker is saying. That's normal. Alternatively you may be in a Zen-like state of peace and calmly looking forward to your talk. That's normal as well, but perhaps a bit less normal.

So now you're due to talk. Take a big breath and walk up to the podium. Hold onto the podium to steady yourself, if there is a podium, or hold onto something, even a pen, and feel its solidity giving you strength. Try to smile at the audience and think warm thoughts towards them. They want you to do well. If you bear this in mind it can help with nerves, and in fact, most of them genuinely *do* want you to do well. Say your first sentence and … you're off!

When you stand up to talk, try to look at the audience *more often than you look at the slides*. Just remember to turn and look at them from time to time as this will help you to project your voice more clearly and to engage better with the audience. And try not to stand in the way so people can't see the screen (Figure 8.1). Don't rush, because they want to hear and understand what you are saying, and if you can, convey enthusiasm, because people will respond to all your positive energy.

Sometimes the results from your research can be completely depressing, it's all gone horribly wrong and you feel ghastly. Well, science can be like that, so just explain, and if anything most of the time you will have an audience that's sympathetic. We've all had disasters in our work and know what it feels like. But take the time to explain to people what happened and present any conclusions that can still be drawn.

If you have a large audience, as you look around, you will see a range of levels of attention. Some people will be listening to you in rapt awe (probably); others, not in your topic area, may actually be doing their email and clearly thinking about whether Spurs are going to beat Manchester United. Do not be put off by the small percentage of people who aren't paying attention — they come with all audiences and everyone who speaks regularly is used to ignoring them. Their lack of interest isn't a statement about you or your talk, it's more of a statement about *them*.

There is a well-known professor in the field of biomedical research who is really quite rude. One of us was at a very small meeting where this person sat at the back of the room and NOISILY turned over all the pages of the local newspaper while people were giving their talks. It was very off putting for new speakers. Ignore this rude behaviour if you ever encounter it.

Use your laser pointer to draw attention to the key points as you discuss them, and to point out important details on figures. Don't wave it about too much as this can be very distracting, and above all never point it directly at the audience!

Try not to flick backwards and forwards through the slides during the presentation as this can be quite disorienting for your audience. If possible, keep to a linear progression through the slides — if you want to refer back to something that was on an earlier slide, you can always include a duplicate to avoid going back to it.

Sometimes we find that the timing goes wrong, even if we have practised our talk several times. In case this happens, it's worth thinking in advance about which slides could be cut out in the event of a timing problem. Cutting out a section of the talk is better than either rushing so fast that people can't follow you or having to stop in midflow before you have got to your conclusions.

If you are presenting at a summer school or similar meeting, allow more time for questions during the talk and try to set a slower pace.

Figure 8.1 Agnieszka presents her work to the audience and looks at them, whereas Clarissa won't look at the audience and finds it difficult to get her message across.

In this situation it is really important not to leave your audience behind as they are trying to learn about your field from scratch. So take your time to explain things clearly and without leaving out steps in the argument. Your audience will thank you for being considerate because it will help them to learn more.

The approach you need to take to summer schools and similar types of talk is quite different from a standard conference talk, and in some ways it will be closer to an undergraduate lecture (except your audience may be more enthusiastic than undergraduates!). For instance, you might want to make use of a whiteboard or blackboard if there is one for any mathematical derivations. It's particularly important to try to gauge whether your audience has got lost by checking their body language. Try to encourage participants to ask questions during the talk to clear up any difficult points.

8.5 The Talk Finishes. Then What?

You finish your talk, and then in small meetings people may ask you questions immediately, or there may be applause and then people ask questions. In this case it can be very confusing knowing where to look. Do you look at the Chair, or the audience? If the Chair is clearly picking out the questioners then go with the Chair's guidance. If they aren't doing this, then you scan the audience and make eye contact with people whose question you wish to answer. The Chair will tell you when time is up and you can go to sit down.

Try not to worry about not being able to answer the questions. You are an expert of your own project, and if your mind goes blank it's fine to say 'sorry, I can't remember but can go through it with you afterwards'.

If you have no questions, don't get downhearted. Even the most brilliant of talks with the most startling results can lead to no questions: *all* of us have experienced this — maybe it's late, people are tired, or maybe it's just before lunch and they're hungry, or ... many other reasons. Often what you find is that people will approach you afterwards to ask what they want to know, especially if time is short and the session is over running.

Just occasionally

While you are talking, a fire alarm will go off, or there'll be a power cut, or (as one of us has experienced) someone in the audience is suddenly taken ill or … any number of unexpected things can happen. If you can keep talking just do so. If it is impossible then look to your Chair for guidance. Obviously if they're running out of the room to avoid the flames then it's probably a good idea to follow them.

Key Points

- Practise!! And then keep practising!
- Say the words out loud and ask your colleagues to critique your talk before you give it.
- Practise to the correct timing.
- Check that your talk works on the conference computer and AV system.
- Take a laser pointer or presenter with you.
- Wear suitable and comfortable clothing.
- Look at your audience as much as possible.
- Keep calm and don't rush.
- Remember to take your computer or USB stick when you finish …

Chapter **9**

Chairing a Session

The key thing to remember about Chairing a session is that everyone, audience and speakers, like a **strong Chair**, so that they know what they are doing. The people in your session are your sheep and you are their shepherd, so it's your responsibility to look after them, even if they have 50 years more experience than you do (Figure 9.1) …

9.1 Before Your Session

So, as with many aspects of being a professional scientist, you have to prepare in advance. Look at the meeting programme before you leave for the conference and find out if you are expected to give a short introduction to the session. Generally this will not be the case, but check anyway. If so, find out how much time you are expected to talk for, and prepare a few slides giving the introduction to the field — referring as much as possible to the areas in which your speakers are likely to talk. This is all a lot trickier of course if you are only asked to Chair the session the day before, as is often the case! But it's always worth checking what you are expected to do as Chair.

When you get to the meeting, you may know the people in your session, or you may not — if you don't then try to work out who they are and introduce yourself. This is often a good way of breaking the ice and networking: playing detective to track down all your speakers. Sometimes people even email the speakers in their session before the meeting, just to say hello and introduce themselves.

Figure 9.1   The Chair is in charge of the flock.

Your next job is to familiarise yourself with how the audio-visual (AV) system works, so that you know where the microphone is when you have to introduce people, and how to open their presentations on the conference computer. If the meeting has professional AV folk then find out where they sit and even say hello to them and ask if all your session speakers have loaded their talks, or if you need to chase anyone to do so. If speakers are going to use their own laptops, make sure that you know how they should be connected to the AV system. Most scientists can think of many occasions when talks have been held up unnecessarily because no-one knew how to connect the laptop to the projector.

If you are in charge of AV, for example at a small meeting, then also make sure you know how the lights work. For some slides, the lights may need to be turned off for particularly dim or subtle images, and you — or someone you ask to do this — are on duty.

Finally, if water is provided, make sure it's within handy reach of the speakers and, if in a bottle, the bottle can be opened easily. One of us recently went to a huge international conference where water was provided in sealed bottles that required a bottle-opener. Unfortunately no one thought to leave out a bottle opener.

9.2 At the Session

Keep the meeting programme to hand, so that you don't have to worry about forgetting the speaker's name or affiliation or subject, when they come up to the stage: simply read it all from the programme. And remember to practise the correct pronunciation of names that are unfamiliar to you — if this is a real problem even ask the speaker what is correct — most people won't take offence and would rather have their name said properly.

There are two reasons we have Chairs, in order to (1) keep to time and (2) run question sessions after the talks. Here are our hot tips about both.

Keeping to time can be very difficult, but you are the only person in the room who can do this job, and everybody in the audience is depending on you. Honestly, they are, they really are.

Let's start with the speakers: people want to talk and often the oldest and most august of us are the least disciplined. But no matter how distinguished the speaker is, it's important to keep them to the allocated time so that you can move on to the discussion and get the session to finish in time for lunch.

So how do you control the session? And **you have to**, it is your professional duty because you're **the Chair**.

Just before the session starts, identify your speakers and remind them of points 1–3 below. If you can't see them all, then you can say points 1–3 as you announce the session title at the beginning of your session. All speakers need you to tell them how much time they have left so they can work out if they need to hurry up or not. And then they need clear directions about when their time is up. Some will be nervous and depending on you to help them (keep smiling and nodding encouragingly at nervous speakers, it helps everyone). Remind the speakers of the following:

(1) How much time they've been allotted for talking and for questions.
(2) That you have a loud timer/will wave at them/dim the lights 5 minutes before their talk is due to end.
(3) That you will stand up/use a louder timer/hold up a flag/turn off the microphone at the time their talk should end.

Most speakers want to finish on time, but simply don't keep track of the minutes, so your job is to do it for them. Remember to bring a timer or flag to wave at the speaker at the allotted time. And remember to wear a watch or have your phone handy in case you cannot see a clock. And finally, the speaker of course has to be able to see you.

One of us many years ago was Chairing a session and thus in charge of timing, with some really verbose speakers. When this person sat down to Chair, they realised the speakers couldn't see them waving that time was up …

If you have someone who REALLY WILL NOT STOP TALKING then a couple of minutes after their finishing time, you have to get up, stand on stage and then ultimately politely ask them to finish, because the conference needs to move on. The audience will be sympathetic, because not everyone will be fascinated by what that particular speaker is saying and many will simply want them to get off stage.

Once the speaker has finished, thank them and start the applause yourself and everyone will clap too (they will, we promise) and move swiftly on to the questions (if you have time).

One of our colleagues uses a high-powered water pistol to get recalcitrant speakers off stage. This is a particularly robust style of Chairing that we do not recommend (Figure 9.2).

Figure 9.2 Sometimes special measures have to be taken to get Professor Karloff off the stage.

9.3 The Question Session

Often, the most interesting discussions come at the end of a presentation, just as the next person should be up on stage talking. So there is a tension between having an interesting discussion, and massively over-running the session. No one minds a session that runs over by 10 minutes. People mind greatly if their lunch is delayed by an hour — and often there are time constraints because catering is only available in specific timeslots.

Several people may want to ask a question and if they all do it together no one gets heard and no one gets answered. So someone has to take charge, either you or the speaker. Before the speaker comes up, tell them whether *you* will field the questions (i.e. you will point out members of the audience to ask questions) or you expect *them* to field the questions, so they make eye contact with audience members and answer directly. If you don't do this, the audience doesn't know whether to look at you or the speaker when trying to catch someone's eye to be given permission to ask their question.

It may be easier if the speaker fields the questions, but sometimes in a dark auditorium, they may not be able to see the audience as well as you do, as a Chair, sitting at the side.

As we mentioned above, often the most interesting points come out in the questions after a talk, but no one likes running 2 hours over and being late for the drinks party. So it can be an uncomfortable decision about when and how to cut short the discussion. In this case, you can simply finish the question session by saying something like *well, there are lots more interesting questions on this topic but I'm afraid in the interests of time we have to move on. I'm sure this discussion can be continued over lunch/in the bar later.*

Equally, if the audience is silent, it is courteous to have a question prepared for the speaker — again difficult if you're not an expert in their field. This is why you need to read the Abstracts of your talks beforehand and have an idea of what to ask. However, if you are really short of time, then you can graciously say *in the interests of time, we need to move on, but I'm sure there are many interesting questions for discussion over lunch/in the bar later ...*

One of us has a colleague who in this situation always asks the speaker what they think about the polarisation effect. In physics there is ALWAYS a polarisation effect and so it's either (1) a sensible and routine question which deserves a thoughtful answer or (2) an amazingly perceptive and difficult question that puts the speaker into shock as they hoped that no one would mention it. Polarisation can mean many different things, so it's perfect for when you have no idea what the person was talking about. If you're a physicist.

What to do when a speaker and a questioner get into a row? When two or more people get into a heated discussion, you have to step in and defuse the situation, and move on. This can be difficult if people are starting to get quite emotional, but the Chair has to handle the situation in, if possible, a humorous and graceful way, or simply by stating the obvious: *that there is no time to continue this interesting discussion but it can go on later ... in the bar* (obviously a lot happens in bars at meetings).

Other crises, outside your control, can occasionally happen such as when there is a power cut, or the building is on fire, or a speaker doesn't turn up, or the conference computer explodes. In each case, remain calm, and do what is most sensible, which might be asking people to leave the building because of the fire alarm/power cut, or moving onto the next speaker. **You** are **the Chair**, and **you** are in charge.

9.4 Types of Talks You May be Asked to Chair

Most people just give contributed talks but occasionally there are special talks from particularly important speakers, or marking particularly important events. These can be Plenary Lectures, Keynote Talks, Named Lectures — so if you see one of these in a programme, what does it mean? Briefly, a Plenary Lecture is a broad talk that is timed to occur when the whole conference is present, and no other parallel sessions are taking place. A Keynote Lecture or Talk tends to be more focussed on a particular theme; usually the Keynote speaker will have been invited so that they can talk in more depth about a specific area. However, as usual with academia, there are no hard boundaries and Plenary and Keynote are being used interchangeably. A Named Lecture is simply that: a lecture given in honour of someone, usually deceased. In addition, there may be Invited Talks, in which a speaker is asked to give a presentation that sets the scene for the following short contributed talks on a particular topic.

9.5 At the End

Simply thank all the speakers and the audience for an interesting and enlightening session. Keep the thanks as short as possible as everyone will want to get out and stretch their legs and drink more coffee/wine/beer. You may thank funders and even the AV folk, but keep it short.

Key Points

- Check what you are expected to do before you get to the meeting.
- Find your speakers before the session and introduce yourself.
- Scope out the AV arrangements before your session.
- Tell your speakers how you are going to signal they are near the end of the talk, how many minutes they will have left, and when the talk should be finished.
- Tell your speakers whether you are fielding questions or they are.
- Have a question prepared for each speaker in case you need it.
- Get all speakers off the stage, even if the discussion is fantastic, in a timely manner and politely; but don't worry too much about over running by 10 minutes or so for the session.
- REMEMBER: EVERYONE LIKES A STRONG, POLITE, PROFESSIONAL CHAIR.

Chapter **10**

Talking to the Public

10.1 Why We Should Talk to the Public

There are many many reasons why we should explain what we do to the public. This process of engagement between scientists and the wider community, now often referred to as *Science and Society*, has gained in importance over recent times. We could write an entire book on *why* the public understanding of science is essential — and it's up to scientists, US, to engage with this process and to enter into dialogue with other members of society. It is also increasingly (and rightly) becoming a career requirement for scientists to engage with the public dissemination of research. We think this is a Very Good Thing, but you have to know how to do it. Most universities and research institutes offer courses in talking to the public, and these usually give you good ideas and some practise. We strongly recommend you to go on such a course if you have the opportunity, and certainly you must have some appropriate training before presenting science to the public.

Much research is funded with public money so remember this when talking to people whose taxes are allowing you to follow your passions. They have a right to know what you are doing and why, and moreover, they have a right to expect you to engage with them in discussion about it, and not just to tell them the way it is. So our involvement with Science and Society activities should be a dialogue, not a one-way transfer of information.

At a meeting, even a scientific meeting, there may be sessions for the lay public, especially if it is a medical research meeting. Or, you may be invited to talk at a science festival, for example, which is all about people without scientific training enjoying and celebrating and understanding science. So here we want to talk about *how* to talk to the public.

10.2 How Should We Go About It?

Everything we do can be described in the utmost complexity, down to the concentrations of the solutions we use, or the angular resolution of the radio telescopes or the non-parametric statistics that might be important for our electrophysiology studies. To explain the complexity we write papers, theses, monographs, books for our peers, and to keep everything as short and readable as possible, we use technical terms that we all understand within our own fields.

However, *everything* we do can also be organised into just a few sentences that give the essentials of an explanation that is understandable by everyone. We recommend again that you use the **why, how, what, why, who** structure (just as we did in Chapter 5):

- **Why** are you doing this research?
- **How** are you going about it?
- **What** have you found out?
- **Why** are your results important?
- **Who** gave you the resources to do it?

For example, Abena Asante is carrying out a PhD on genome editing of mice to create new models to study hippocampal degeneration in the CA1 region arising from amyloid plaque deposition that replicates what is seen in human Alzheimer's disease. Got all that? She has a very nice poster she takes to scientific meetings, she's writing a lovely thesis, she has a good paper published, it's all going very well. But her dad's coming to town and is going to take her out to dinner, and now she has to explain what she's been researching for 3 years, to her dad,

Figure 10.1 Abena and her dad, Mr Asante, waiting for their puddings.

who has a degree in Modern Russian Literature, has spent almost all his working life in a large clothing retail company dealing with accounts, and who's secretly a bit frightened of science (Figure 10.1).

Abena says just five sentences:

'Dad, [**why**] Alzheimer's disease is really common, and we don't know how to treat it or really what goes wrong in detail in the brain. [**how**] I'm making genetic changes in mice to replicate some of the features of human Alzheimer's disease, so that we can study the effects in the brain in a way we can't do in humans. [**what**] I've found out from my mice and work that other people have done, that a particular protein collects in a part of the brain that's important for memory and that's a bad thing. [**why**] These findings are going to help us understand Alzheimer's disease and, ultimately, how to treat it. [**who**] I have good funding from a medical charity, and my PhD supervisor, Professor Karloff, is fantastic'.

Abena's dad understands as much as he wants to about what she's doing and why. He can see the point of the research and its importance, and he can see that it is necessary for a better future. He's proud of Abena and he feels engaged by what she has said, and he can explain it to his friend Maurice when they meet next. All that in just five sentences. Abena and her dad drink a very nice Muscat with their pudding and decide to go on to a cheese course.

10.3 Know Your Audience

Anything complex can be simplified to the appropriate level for conveying the information to a particular audience. And the first step is to understand your audience. Do they have any scientific background, perhaps in another field? Or are they a lay audience in which most people have no, or very little, science? When talking to the public you have to assume no science knowledge, and explain the basics of your research entirely using normal words, or, if you have to use the occasional scientific word, then carefully explain what that word means. That doesn't mean your audience is stupid. They may be super clever and highly motivated to learn about your topic. But it does mean you have to think carefully about the words and concepts you use, so that you're not having to pause and explain every single technical point, which would make the talk last several days …

Even when talking to undergraduates about your research, perhaps in an informal talk to an undergraduate society, you have to be careful about assuming too much knowledge. An undergraduate just getting to grips with the basics of their subject is going to struggle if you jump straight into advanced research concepts. So remember that

Figure 10.2 The local Science Society listen to Professor Karloff's Public-Understanding-of-Science talk on *The Quantum Relativity of Landfish Parapsychology,* with equations.

many of the ideas we discuss here will apply to this sort of event as well as one involving people with no scientific knowledge at all (Figure 10.2).

10.4 Structuring Your Talk

As with the advice about scientific talks, check how long you have to talk for and then plan your talk according to the **why, how, what, why, who** structure. Think about what the audience really wants to know, because the emphasis of your talk will be a little different from a scientific talk to your peers.

The general audience wants an introduction that starts with The Big Picture, sets the scene and relates what you do back to everyday life.

The '**why**'. For biomedical research this may be obvious: perhaps you are researching a particular disease, or trying to improve the crop yield of a cereal plant. Or perhaps you're a climatologist and the audience is worried about the effects of climate change. At first, some subjects may seem too esoteric to talk about, but … they're not! Just think about The Big Picture. Cosmology is hard to understand in detail unless you're a physicist, but everyone is fascinated by it because it tells us about the story of the universe, which is the place we all live in. Computer science may seem hard but it has led to artificial intelligence and robotics, and we will all be affected by those topics. Engage and enthuse your audience. Explain why what you are doing is relevant to them — philosophically and/or practically.

Then with as few big words as possible, explain the bones of your project, the '**how**'. Tell a story and take the audience with you. Make the story comprehensible by simplifying and explaining carefully and do NOT use technical terms. Even if you try to explain technical terms, you will lose a large portion of the audience because either they'll forget what you mean, or they'll be frightened that they won't understand it all and will zone out. So no technical terms please. Rather than say 'I used photoionisation spectroscopy based on the absorption of electromagnetic radiation with a wavelength of 422 nanometre to carry out an isotopic analysis,' you could simply say 'I shone a laser through my sample and looked at its effects so I could measure what was in the sample.'

You are an adventurer leading them confidently into your world. Use loose analogy if that helps. 'Enzymes are macromolecular catalysts such that that less free energy is required for the substrate to react to form a new product,' might be re-stated as 'Enzymes are some of the machinery of cells, they do things'.

The '**what**' in this context, is what you found out. And again, keep it simple: big headlines, and use analogy if helpful, but no technical terms. Your audience are not idiots, they're listening to you, voluntarily, because they are interested and want to learn more. So give them what they want: help. Help to understand your world.

For the last part of the talk, pull back to the bigger picture, reminding the audience of what you said in your introductory session — **why** what you have found out is interesting — and where it's going in the future. Finish by thanking the funders and others **who** have given you the chance to talk and/or enabled you to do the work. Especially if they're in the audience.

10.5 Creating Your Slides

It's possible that you won't be able to use slides for your talk, depending on the type of event it is, but in many cases you will have the possibility to show some slides and generally it's a good thing to take advantage of that. But be wary of reusing slides from your technical talks as they may not be comprehensible to a non-specialist audience. You may well have to spend time creating entirely new slides, with a stronger focus on visual appearance and accessibility than for talks to fellow scientists. In this case it may well be appropriate to (selectively) use more animations in your slides.

Slides for scientific talks in general should simply show the data and information you are trying to convey. As a general principle you should always try to keep the number of slides small and keep the amount of text on each slide low. This is also true for presentations to the lay public but ... even more so. Keep it simple:

- No equations, absolutely no equations. *Ever.* You will have lost most of your audience as soon as they spot an equation. Try using words instead. Think about science programmes on television: you almost never see an equation.
- No complicated graphs. Most people are not used to looking at graphs in detail, so if you have to use them, present them in the simplest way, with colours, and only use a graph to convey one key point. Newspapers will only use clear and simple graphs like this.
- Use diagrams where they help explain something. Again, think of TV programmes or newspapers. They will use simple graphics to get something across which otherwise would need lots of words. Many of us think visually so a diagram can be an effective way to make information accessible.
- Sometimes simply showing a photograph may be helpful, but chose such pictures carefully so they are clear and they do not offend or upset.

Remember that your audience is going to be worried they won't understand and so they want you to lead them by the hand. They won't have the capacity to follow what you are saying **and** work out what's going on, on the screen, if it's remotely complicated (none of us do outside our own specialities).

Finally, your slides also need to convey enthusiasm, so this is probably the only time when we will ever say that you can use exclamation marks ... if you want to!

10.6 Important Things to Remember

Be sensitive to your topic

Whether you work in medical research, or on the geochemistry of fracking, there are people who may be greatly affected by your topic and you have to respect their experience, so no off-hand jokes or sarcastic remarks. People will be ultra-sensitive to anything that affects them directly: all of us are.

Always present your topic in an appropriate manner and be sensitive to the words you use — this is particularly true for talks in the medical arena, where it is especially important to talk with experienced senior colleagues about how to present your work, as the words you use may have huge significance and impact for people in the audience who are directly affected by the disorder you study. Remember also that what you say may be videoed nowadays and will almost certainly appear on social media of some description, for ever.

One of us was giving a genetics lecture at a Natural History museum many years ago, and briefly got onto the topic of mutation and cancer, mentioning a rare cancer, Ewing's sarcoma, as an example. After the talk, a boy in the audience came up and said he had Ewing's sarcoma. This is a very very serious disorder. It was discussed sensibly by the speaker during the talk, and the boy wanted to chat more afterwards, so the experience was rewarding for the boy and as it turned out, very rewarding for the speaker.

Funding can be a sensitive issue for some people. They will be aware that huge sums of money go into scientific research and they may feel that money spent on giant particle colliders or space tele-scopes could be better spent on trying to solve pressing social prob-lems. So be prepared to point out the benefits of fundamental research if this is your area, and explain how future applications often build on fundamental advances in basic science that was not expected to produce applications when it was carried out. For example, an eso-teric topic like Nuclear Magnetic Resonance in atoms and molecules would not originally have been expected to lead to the massive appli-cation it has now found in medical diagnostics with MRI scanners. And remember that in many cases our funding ultimately comes from the taxpayer, so people have a right to know why research is worth funding. It is interesting to note that the UK Academy of Medical Sciences states that (at least in the context of cancer and musculoskel-etal diseases) research has shown that "every £1 invested in medical research delivers a return of 25p per year, *forever.*"

Statistics

Many people are suspicious of sophisticated statistics because they are abused in so many contexts, so we have to be careful in the way we use them. Of course you should show how your conclusions are justi-fied by statistical data but try to make the presentation of any statistics as clear and straightforward as possible to that it can be understood. You don't want people to think that you are manipulating the data to make sure a particular result is found.

Back in 2009, many emails sent by climate scientists in the UK to each other were leaked to the press. They were accused of manipulating the presentation of climate data to make sure that it was done in a manner that clearly supported the generally-accepted view of climate change. This did severe damage to the reputation and integrity of climate scien-tists and demonstrates the dangers of being accused of misusing statistics.

So try to avoid detailed statistics and technical terms where possible. For instance, "the spread of the results" might be better than "the standard deviation of the results" when talking about sets of measurements. Terms like "linear regression" could be intimidating.

Don't present lots of numbers as most people would find it difficult to take that in. As we said above: keep the message simple.

Be responsible

Most people are not as lucky as you are and haven't had your scientific education. So even if you are at the bottom of the academic ladder, you're the expert and a lay audience may take you words to heart, and make decisions, such as whether to vaccinate their child, based (partly) on what they hear from you. We don't want to frighten you: you are not responsible for other people's decisions, but you are responsible for what you say. So give facts, and be careful with opinions, and if you are asked questions it's fine to say that you don't know the answer.

Difficult topics

There are many potentially difficult topics that can crop up in talking to the public. Animal research is an obvious one. For difficult topics, take advice from people who you know are regular communicators to the public, and whose views you trust and respect, and who have integrity. For example, one of us works on animal research, and always acknowledges that in any audience there will be a range of views, and that if we had alternatives we would work with them, instead of animals. So as far as possible (in the time available) the topic is acknowledged and alternative views considered.

Similarly, climate change can be a very sensitive and emotive topic. There is a lot of misinformation about and there is huge potential for your words to be misinterpreted or misrepresented by others. Try to be balanced as much as possible and stick to the facts. The same ideas hold for areas like nuclear power as well.

Rehearse your talk to non-specialists

As a sanity check, give your talk to your friends who are not scientists. Check that they really have understood what you're trying to convey and why it is important. If not, make changes according to their suggestions.

Posters

Occasionally you may need to present your work to the public in poster form. Here, the same guidelines as above apply.

Remember

People love science and some topics can really capture people's imagination. Science tells us about ourselves. The word itself, science, comes from the Latin word (*scientia*) for 'knowledge'. General science festivals attract huge audiences and scientists on television often become national heroes (as they should, of course …). So do not be afraid of your audience, but you have to respect that they are there to learn, and you are there to teach them. So teach — convey what you are doing and why, and try to convey also the dynamism of science and other aspects such as how international it is. How we are all part of a global network of people researching the universe about us and within us. But also listen, because the people you are talking to also have valid opinions that should be heard and appreciated. Be enthusiastic and let it show that you are excited by your work. Actually, discussing science with the public is a great job!

Key Points

- No technical terms.
- Concentrate on simple concepts well explained, with analogy as appropriate.
- Be sensitive to the words you use and how the topic you are studying may affect people in your audience, or any audience.
- Be careful in the way that you use statistics.
- Try to enter into a genuine two-way conversation with your audience.
- Be enthusiastic!

Chapter **11**

Practical Arrangements for Going to Meetings

So, you've found the perfect conference that you want to go to, to present your results, meet like-minded folk, and generally have a good time, or at least visit a new location. How do you get to where you want to be?

11.1 Before

Register for the conference, through the online website. Some conferences are competitive, and attendees are selected from the Abstracts they write; these tend to be the smaller meetings or workshops. Here, just because you register doesn't mean you will be accepted, so write your Abstract carefully so it is likely to be of interest to people at the meeting. Some meetings accept you as soon as you register, and they will want payment immediately for registration fees. These fees cover the cost of administration of the conference and hire of the venue. They often cover some food and drink too, but check exactly what you are paying for, as the all-important Conference Dinner (at which elderly scientists reveal that they are spectacularly good dancers, or not ...) may require an extra payment, and sometimes there are cultural trips that you may want to take too (Figure 11.1).

How are you going to pay for the conference? Your supervisor or Department may have the money, but otherwise look around for grants. Many Universities have internal grants for travel, but also you

Figure 11.1   Professor Karloff enjoyed the dance after the conference dinner.

should be a member of a professional society — in the UK these are known as Learned Societies, which is a rather archaic term for describing the professional body that represents your scientific interests. All countries have them and they often give out travel grants for members, especially students and junior scientists, to go to conferences, sometimes in exchange for a short conference report for their own journals. Other funders may also give out travel awards, so ask around and look on the internet. In addition, the conference itself may give out travel awards to junior attendees, so have a look at the meeting website, and start applying. These travel awards also go onto your CV, showing that you have been successful at competitively applying for funding.

The next two issues are accommodation and travel. Accommodation may come with the registration fees, or not. The meeting website may direct you to specific hotels, in which case book early to get what you want. For big, popular meetings, you may be able to get together with friends/colleagues and hire an apartment for yourselves, which can be cheaper than hotel rooms and fun if you can work out the finances and the timing. All the information you need should be on the website and remember that for the absolutely huge meetings, where, for example, over 30,000 people descend on one city, the rooms nearer to the conference centre will go very quickly, so book well in advance.

Travel also works best if booked well in advance so you can get cheaper seats. For long-haul destinations the absolute cheapest way to get from your home to the conference may be via Pluto with multiple stops on other passing hub planets. Try to avoid this. You will be tired enough from the jet-lag, so if you or your supervisor can afford the flight, always go for a direct flight or at least the minimum number of stops/changes. Remember that your Institution may require you to book through a particular travel agent and this may complicate and/or delay getting your flights organised.

Don't forget that for some countries, such as the USA, you absolutely HAVE to have Travel Insurance — your University or Institute should be able to book this for you, or it may well have its own group travel insurance policy that will cover you so long as you register your trip before you leave home. Do not set out on your trip without proper insurance.

One of us, many years ago, knew a European postdoc who went to a conference in USA, and partied quite hard. As a result they broke their leg ... and didn't have any travel insurance. It took years of the postdoc's salary to pay for the hospital treatment ...

Now you have registered and been accepted, you have booked and paid for your room and your tickets. What next? Get a visa well in advance if you need one (and if you're not sure, then check immediately), get your poster and/or talk ready and in a format that's easily transported, remember all the usual things you take with you as well (medication, swimming apparel, sunglasses, music, laptop, mobile phone, PASSPORT AND TICKET AND CREDIT CARD), and get ready to go! And, do one other thing: keep your travel receipts safe, if you need to claim the costs back.

11.2 During

Once you've found where the conference is, and your accommodation, then there should be almost nothing organisational for you to worry about, but life happens, and it may be that you need to dash home urgently or you get arrested or you get ill. All meetings have organisers, and the organisers have a responsibility to help you, so seek them out, and ask for help. Most of the time (not always) they're locals and will know what to do. Occasionally they're also visitors but they will know the local staff for the venue.

11.3 After

There may be a few jobs to carry out after the conference, aside from emailing the nice postdoc you met at the Gala Dinner. These include sorting all your receipts into a tidy pile and claiming any travel and subsistence expenses that you are entitled to. Do this immediately, even though it's a boring job, because it won't get any less boring if you leave it, you will forget what each receipt is for, especially if it's in a language you don't speak, and because most Universities/Institutes have time limits on expense claims, so if you don't claim within 3 months, say, then you can't claim at all.

One of us is really bad at claiming expenses promptly and had to make a special case to the Finance Officer to explain the reason why a claim for over £1000 was more than 6 months late (the problem was that there was no good reason). Next time it's unlikely there will be a happy outcome (Figure 11.2).

You may also want to follow-up on possible collaborations or interesting points. So as well as emailing the nice postdoc, get in touch with the other potential collaborators you met and initiate useful conversations about where to go next with your work. You may also be required to give a report of the highlights of the meeting at your own lab meeting or an organisation that gave you a travel grant, in which case do this immediately while you fully remember the significance of everything.

Figure 11.2 Don't forget to sort out your receipts in good time.

Then relax. With any luck, even if you are the shyest person in the world, it wasn't all ghastly, you made a good presentation about your work, heard some interesting science, met nice people, saw a new place. You are a scientist who has been to a conference: you are a professional.

11.4 Social Media

Social media come in many shapes and forms from LinkedIn to Facebook to Tinder and beyond. While some meetings will have their own Facebook page, the most useful social medium currently (and we know this will go out of date as soon as we put the words on the paper) is Twitter, because it is designed for real time updates, so you can instantaneously know what's going on and comment on it.

The organisers of the event will usually publicise (on their website) the Twitter hashtag for the event, and you can then start to follow the meeting as it develops.

At the meeting, Twitter can be a useful medium of quiet background conversation for the organisers *this room is too hot, I'm at the back and I can't hear the speaker*, but the main excitement comes — or should come — from what's being presented *Professor Karloff has just given an extraordinary talk about Moog Genesis — lots of questions from the Parapsychologists! Professor Karloff appears to be hitting one of the physicists. Ouch! The police have been called.*

As with all human interactions, please respect the rules: if you are asked not to Tweet about, write about, photograph or record, a speaker or a talk, then please respect the request, and don't do it. Most meeting organisers now assume that people *will* freely send out information, so when you're specifically asked *not* to, please don't.

Following someone's Twitter feed is also a useful way of making connections, especially at huge meetings where you may not know what the person looks like, or stand a chance of bumping into them, but you would like to meet anyway.

Key Points

- Register, book your accommodation if you need to, and book your travel IN ADVANCE. Work out how you are going to pay for it all.
- Book yourself into the Gala Dinner as although they can be excruciating, especially if the organisers put on an entertainment that requires audience participation, it's often the best time for meeting and chatting to people, when everyone is more relaxed than at the scientific sessions.
- Apply for travel funding if you are eligible.

- If you have any problems during the meeting, go straight to the organisers.
- Do your travel expenses and any other chores immediately after the meeting.
- Follow-up on contacts you made at the meeting.

Chapter 12

Making the Most of Your Meeting

One of us doesn't particularly like going to conferences (it's a confidence thing probably) and is constantly amazed by the enthusiasm of other people for attending them. This author moans about having to go, and then usually (but not absolutely always) has a great time, especially if the meeting is small. We're all different, and you may just leap into a conference like a dog launching itself into the sea; or you might be more like one of the authors — a snail only slowly sticking its head out of the shell. Whatever your confidence/style/enthusiasms, here are some points to think about beforehand.

12.1 People

Amongst the people attending the conference will be people like you: extrovert or shy, confident or unconfident. Whatever the external appearance, we all have similar issues. If your conference lasted for a year, we'd all make friends with someone there simply because we had more time — so as most conferences are short, you might as well go to the social events, even for a few minutes, and try to network, because in fact if you stay in roughly the same field you will see the same people year in year out anyway. Sometimes that can be a consoling thought for the shy: conferences get better, the more you go to.

There will always be a mixture of people from senior academics to starting PhD students. So you don't need to feel out of place just because you are a relatively junior person (in fact, often the majority of people at a conference are PhD students or junior postdoctoral researchers). The conference is for you just as much as it is for other more senior attendees. We all come with different needs, but conferences are set up so that everyone can benefit from attending.

Incidentally, when we say 'try to network' this doesn't mean you have to talk in depth, 100% of the time, about science, fun though that is. Often people really appreciate someone who can talk generally and just break the ice by making conversation. A cliché about the British (for example) is that we talk about the weather. This is sadly true, as the topic of RAIN is often in our minds, but it's also a topic that everyone can join in with. It sounds facile, but a simple observation about the weather is often a good opening to discussing where people come from and that leads on to many other topics, geographical, political, personal, etc.

If you're a more confident type, then take the plunge and introduce yourself. Shyer types will appreciate it, because it opens the door for them to do likewise. You'll get to know lots of interesting people this way. Rarely, people take this too far and you need to be sensitive to the fact that some people may not be ready to talk to someone they don't know. Some of us have met people at conferences who go over the top in forcing people into conversation and making sure that everyone knows who they are.

If you go to a conference with a bunch of people from your own group, that can be great fun, but then also try to meet new folk and to be generous to people on their own — invite them along. That person on their own might be you at the next conference. Or might turn into a lifelong friend. The easy option is to stay with people that you already know, but it's always worth making the effort to get to know new people. Try it.

12.2 The Venue

Regardless of how exciting the science is, how wonderful the people are, how great the wine is, the conference venue does make a difference. A fabulous setting makes us all feel more fabulous. A really really down-market setting can be depressing. For those sensitive souls, such as one of the authors, you can acknowledge this and if you're stuck for a week in a concrete bedroom with a glaring strip light, sharing with someone you don't know and wouldn't choose to know, then it's not great. So help yourself to feel better and that will help your confidence. Buy some flowers for your bedside table. Or a souvenir that makes you laugh. Skype your best friend. Help yourself to have a more open state of mind.

Also don't miss the opportunity to see a new place. The programme may not give you much time to look around but if you can, try to get away to take advantage of the opportunity to explore a location that you have never been to before. You might even consider staying on for an extra day or two after the meeting has finished so that you can take more time to explore.

12.3 Preparation for the Meeting

Before you go, think about the topic area and to make the conference even more interesting, do a bit of basic reading beyond your own research. You will also come across the names of scientists in the area, and it's always good to know people's names — especially if you are thinking about possible jobs later and want to relate the research to the person.

When you travel, especially if it's a big conference, you might see people who are likely attendees, perhaps they're carrying posters. Reading the latest issue of *Nature* or *Science* also gives a bit of a hint that someone is a travelling scientist! If you can, strike up conversation briefly — you'll probably bump into them again at the meeting, even if it's huge. It's nice to feel part of something larger and that you know some people, even if it's only slightly. Equally, this means people will remember you. So if your normal travel wear is a zebra patterned onesie, that's absolutely fine, but that's how you'll be remembered.

12.4 When You Arrive

Depending on the time of day or night, you may go straight to your hotel and get some sleep, or you may go straight to the meeting to register. When you register, it may be super busy and chaotic, or running like clockwork. There are lots of variables but we want you to feel comfortable. So give yourself plenty of time to register and keep a sense of humour and calmness if people around you are feeling the stress. Take care of your body too — it's all yours. Conferences don't always provide coffee/tea/water so at least make sure on day 1 that you have what you need. Buy a bottle of water at the airport/rail station when you arrive if you want to.

12.5 Breakfast

One of the authors made the choice many years ago that it was better to grab as much time as possible in bed, than to get up for breakfast. And also had the realization that breakfast is really important. So what to do? This author takes dried or fresh fruit to meetings, so that there is always something to eat, especially if the venue provides nothing till lunch time. This author also takes a plentiful supply of painkillers to counter the effects of conference drinks receptions. But that's another story.

The other author is more keen on a proper breakfast but does get a bit stressed about whether there will be someone to sit with or not. But this is silly really — if you do sit on your own it is most likely that someone will join you. After all, they probably feel the same way about sitting on their own.

Look after your body, make sure it is well watered, fed, slept and comfortable and that will help your mind to feel that way too. You may find that you need to take breaks from the conference hall to get some fresh air from time to time.

12.6 Stamina

Conferences require stamina. Organisers have to cram as much as possible into each session, usually with limited time for breaks. Then there are meals, and then usually nightlife or even skiing or sports in some venues. It can be completely relentless, but like anything else, the more you put in, the more you get out. One of the authors doesn't ski, is not sporting, and never signs up for any extracurricular activities, but somehow, in a conference with lots of people around, especially in smaller meetings or more isolated venues, the time flies by. But it only ever lasts a few days, and the experiences and bonding can last a lifetime. So be brave, throw yourself into it.

Conference programmes are often very crowded and can be pretty exhausting. But there is a limit to how many talks you can go to in one day while both staying awake and concentrating. One of us has been guilty of falling asleep at a conference where the seats were far too comfortable and the lights had been turned down for a talk …

So if you find that you are not able to concentrate any more, it's best to acknowledge it: take a break and get some fresh air. Then you will be able to concentrate much better when you return. Knowing that this is likely to happen at some point, we recommend that you plan which talks you really want to go to in advance so that you can build in breaks. This will ensure that you are really alert for the talks that are most important for you.

12.7 The People You *Have* to Meet

When you look through the conference programme — as of course you should do as soon as possible — there will probably be people you want to meet. Perhaps a future boss. Perhaps a postdoc who can tell you about a lab you're interested in. Perhaps a future collaborator. Perhaps your cousin's husband's cousin's aunty whom you said you'd say hello to. Perhaps someone you once co-authored a paper with, whom you'd like to finally meet in person.

Give yourself the challenge of saying hello to this person and explaining why you are saying hello. Ask the organizers to point out the person if they can, and then choose your moment to go up and say something like, 'Hi, I just wanted to say hello because …'. There. You've done it. If they are super busy and you really need time with them, then ask if you could talk to them over a coffee or tea break. Most people are happy to do this, providing it's for a limited time in a break.

If you want to show your person some data, then come prepared. Take it with you on a computer, or even bring a printout of the data so they can have a good look. Don't waste time trying to describe it in detail in words if a figure or graph is more eloquent.

Key Points

- Make an effort to be friendly to other people.
- Enjoy the location.
- Look after yourself.
- Take a break when you need to.

Chapter 13

Publishing Your Conference Paper

13.1 Introduction

One of the most important aspects of being an academic is publishing the results of your work. In the sciences this is mostly done through scientific journals. After an article is submitted to a journal, it is sent out to referees to be reviewed. The referees are independent scientists and the Editor of the journal makes his or her decision about whether to publish the article on the basis of the reports written by the referees. The Editor may ask for the article to be revised before making a final decision about whether to accept it for publication.

Publishing your articles in the appropriate journals gives them good visibility in your scientific community and demonstrates your achievements. This is important for several reasons: the main one is to make the results of your research available to other scientists who may want to make use of them in their own work, perhaps because they need the results of measurements that you have made, or perhaps because the methods you have used could also be used by them. However, there are other reasons too: to demonstrate to your funding sources that you have used their money wisely; to show your Department that you have the ability to perform independent research that is recognised by other scientists in your field, which is important when you are being considered for promotion; and to give evidence of your track record to potential employers. In the UK there is an additional reason for research to be published in high-quality journals: The Research Excellence Framework (REF), under which all research in UK Universities is evaluated every few years, requires the results of the research to be publicly available, and journal publications are the most effective way of achieving this.

Some conferences publish their proceedings after the conference has taken place. The proceedings are a collection of written versions of the talks and posters that were presented at the conference. Publication of your presentation in the proceedings is usually optional. In this chapter we consider the advantages and disadvantages of publishing your results in conference proceedings. Although there are some general points to consider, the final decision will need to be taken by you taking into account your particular circumstances. The most important thing is that you do whatever is best for you and for your research.

13.2 Types of Conference Publications

Different conferences have different approaches to publishing proceedings. Here we will look at each in turn.

Special issue of a journal

Some conferences arrange with a relevant journal to publish a special issue on the topic of the conference, with an invitation to all the people who presented at the conference to submit an article to the special issue. If this happens, the journal will want to apply its normal standards so there will be the usual peer review process as for other articles. The journal may well appoint the conference organisers as editors of the special issue — especially as the chances are that the special issue will cover a broader range of topics than the journal usually does. These editors will then be responsible for the choice of referees and for making provisional decisions about accepting or rejecting manuscripts.

Typically there will be a deadline by which you have to submit your article to the special issue in order for it to be published along with all the other articles from the conference. This may be a few months after the conference so that the proceedings can be published within a reasonable time, maybe 6 months later.

In this situation your article will be indexed in the usual way (e.g. in *Scopus*, *PubMed* and *Web of Science*). It will therefore show up in searches for articles and the citations will be properly recorded. It will count as a normal peer-reviewed article.

There are some other advantages in publishing your article in this way. First of all, it will be published alongside others from the conference, so it will probably be rather more visible to others in your field than it would be if it was published on its own. Once people know that there is a special issue, especially if it's linked to a conference, they are likely to look through the articles more carefully than other issues of the journal. Also it's likely that the complete set of articles in the special issue will be sent in hard copy to all the attendees of the conference, so this means that they will have easy access to it.

Disadvantages might be that the journal used is not the one that you would choose to publish your work in if you had the choice: it might be not as prestigious as the journals you normally like to publish in, or it might not be a usual journal for articles on your particular topic.

Conference series book

Some publishers publish series of books dedicated to proceedings of conferences in particular subject areas (e.g. Institute of Physics Publishing publishes *Journal of Physics Conference Series* which now has around 1000 volumes). Publications in these books will be indexed so that they can be found on websites like *Web of Science* and *Google Scholar*. Their citations will also be recorded on these websites.

The editors of a conference series book will generally arrange for articles to be refereed. However, the rigour of the refereeing process is variable, although the publisher of the series will have some minimum standards that will have to be followed. For some conferences, the refereeing may be a very light touch procedure so that very few articles are rejected (except any that are clearly not sensible). For other conferences the refereeing may be just as rigorous and tough as for a journal.

These publications have the advantage that all the papers coming out of a particular conference will be found in one place (e.g. on your bookshelf!) and it's likely that many people in the field will have a copy. Many University libraries will also automatically buy books in a conference series. However, in most fields a paper in a conference series like this may not be seen as being as important as one in a normal journal. There are exceptions to this, particularly in fields where progress is very rapid. An example of a field where conference publications are as important as journal publications is Computer Science.

Conference book

The organisers of some conferences make their own arrangements with a publisher to publish a book of proceedings from a conference, but not necessarily in an established conference series. They may prefer this because they can have it published at a cheaper price, or with less interference from the publisher, for example. In any case, for you this shares some of the advantages of a conference series book in that the papers will all be grouped together in one place and most conference attendees will end up with a physical copy of the book on their bookshelf.

However, in this case the book may not end up being properly indexed and so any citations may not be easily found. For instance, one of us looked up our article published in a recent book resulting from a conference. The article was found by *Google Scholar* but did not appear in *Web of Science*, *PubMed* or *Scopus*. So the visibility of your article may not be good if you publish it this way. The standard of the refereeing is also less predictable if it is solely down to the conference organisers.

However, although publication in a book like this may not be the best place for your high-impact research, it can be a good way of establishing some exposure for your work so that there is a record of your findings to date in the public sphere. It's generally fine to publish a progress report in this situation, so you are not necessarily losing the possibility of writing a higher profile publication at a later date. It will also act as a good record of what you have reported at the conference for the benefit of other people who were there.

Summer school proceedings

Summer schools serve a different purpose from normal scientific conferences, as they include a strong education component. Similarly, any publication arising from a summer school is different from other conference publications and will be used in a different way. A well-structured book based on presentations at a summer school can effectively become a textbook for the field. These publications may not end up being highly cited but they can be of great benefit to people entering the field, and they can be quite influential for that reason.

So if you are a speaker at a summer school (or equivalent meeting), there are different considerations to take into account when it comes to deciding whether to contribute to a publication. It may be a lot of work for you if you do a good job, but on the other hand it's likely to be much appreciated by your colleagues and especially by new entrants into the field.

13.3 Should You Publish?

Before making a decision about whether to publish in the proceedings of a conference, there are a number of considerations to take into account. It's hard to give general rules because the standing of conference publications is so different across different research fields, but it's important that you make a rational decision. On the one hand this conference might give you an excellent opportunity to publish in a high-profile book alongside authors that are the top people in your field; on the other hand you might end up spending a lot of time preparing a publication that is never seen by anyone.

Publications serve different purposes at different times. For example, you may be looking to present your latest research results and want them to be seen by the top people in your field. Or maybe you have developed a new technique and you want as wide an audience as possible to hear about it because you think it may have relevance across different fields. We suggest that in most cases this probably means that a conference publication is not the best place to publish this work. You would probably do better to aim for a well-respected journal that is seen by the right people.

On the other hand, you may not have many new results to present but you do want to show that you are working in this field and indicate what you are trying to achieve in your research. Or you may have important new results that are not yet complete, but it would be good for you to stake a claim to obtaining these results before someone else does. In that case publication of preliminary results in a conference proceedings of some sort may be a good way to get on record the fact that you have obtained the first results with a new technique, for example. But you should be careful that this does not preclude you publishing a higher-impact article when your results are complete.

Consideration of how easy your article will be to access is important, as we discussed above. Special issues of reputable journals will have wide circulation and will be found by the major academic search engines. However, a conference book may not be indexed and it may not even go to all the conference attendees if the conference organisers have not agreed with the publisher to automatically send a copy of the book to everyone who went to the conference.

The standing of an article is also affected by how hard it was to get it published, i.e. the type of refereeing that is required. If your results are really good, then you will be pleased that the refereeing is going to be robust and tough to get through. On the other hand, if you just want to get a progress report out there, you might be grateful that it won't be looked at too critically by a referee.

Finally, you need to consider your limited time and whether preparing a publication for the conference proceedings is a good use of your time for the degree of benefit that you will get out of it. Only you can be the judge of that.

Key Points

- Do you have results that are ready for publication?
- Are the proceedings of the conference going to be accessible for the people you need to see your work?
- Is it important to you to get your progress to date into the public domain?
- Is it better to wait to get your results into a journal with higher profile?
- Do you have the time to write an article for the conference proceedings?

Chapter 14

Organising Your Own Meeting

14.1 Introduction

All scientists go to conferences at some stage, whether the main reason is to publicise their own work or to update themselves about other people's work. Most of us derive huge benefit from attending meetings of different sorts.

It's inevitable that at some stage someone will ask YOU to organise a meeting or a conference. Your first reaction may well be to agree, because it would probably be an interesting thing to do. On the other hand, your initial reaction may well be that this is to be avoided at all costs, because it will be too much work.

Our advice it to think carefully about when to say 'yes' and when to say 'no'. Organising conferences IS a lot of work, but it can be very rewarding and it will always be appreciated by your colleagues if you organise a successful meeting. It's worth remembering that we all derive great benefit from attending other people's conferences, so it's not unreasonable for us to be expected to take our turn. But it's best not to agree without thinking through the consequences carefully in terms of additional workload.

As you will have seen from the rest of this book, meetings come in all shapes and sizes, from two to 32,000 attendees. And sometimes more. But they all have certain basic rules for their organisation: people need to be comfortable and looked after, and you need to make communication as easy as possible. How you go about doing this depends on the size of the meeting, so we have divided this chapter according to meeting size, but meeting organisation really comes down to:

- **Communication:** This means informing people in advance that you're having the meeting, informing people about the venue and enabling speakers to inform the audience.
- **Venue:** This means finding the right place for your meeting to take place.
- **Catering:** Catering means exactly that we all work better with cake.
- **Scientific content:** This is what the meeting is all about! But it also needs planning and unless the other things are in place the scientific aims won't be met.

14.2 Small Meetings Such as Lab Meetings

You may have been given the job of organising one, or a series of lab meetings, usually for a few to perhaps 20 people. Here is our checklist to help you:

The venue: Choose an appropriate room. It needs to be large enough for your audience to sit comfortably, and move around. It needs to be convenient, not in the next town. The room may need audio-visual facilities such as a projector, or even a laptop. You may need to book the room in advance (ask your Administrator who is in charge of room bookings if this is the case) and if you're not sure how the projector works you may need to book one of your audiovisual support people to show you what to do, at least for the first meeting. Finally, we all work better in lovely surroundings, so choose the nicest room you can, not the store cupboard in the basement.

The announcement: If this is a new meeting or it's in a different place from usual, make sure that everyone who needs to know about it does know about it. With a small group it's easiest simply to email everyone, but do this at least a month in advance so busy people have it firmly in their diaries. Then email everyone the week before, then the day before, and then on the day of the meeting with a reminder. Some people will still forget what's on or what time, or what venue. We're all busy.

You need to tell people what the meeting is for, which day and time it is taking place, and which room. Don't forget to plan what will take place, with a formal timetable for the meeting if necessary.

The catering: If it's a lab meeting, people may just show up with coffee themselves. But everyone appreciates a bit of home baking or even some nice fruit, so you could ask if people are willing to go onto a rota to bring something to the meeting for everyone. Or you could take the plunge and bake and decorate an entire selection of cupcakes yourself. Go on … scientists are generally reasonable cooks so try experimenting!

If it's a formal meeting with outside people, for example, you may want to ask your Administrator if there is any money to buy in water, cups, coffee, tea, milk, sugar, biscuits. Then you can have those prepared beforehand.

The scientific content: Make a list of who is going to talk. Tell everyone how long they have been allocated to speak for. Leave lots of time for questions and discussion. Send a draft timetable round and ask if anyone wants to make any changes.

On the day, get to the room early: you may need to unlock it and get it ready. You don't want people standing in the corridor waiting. Lay out your beautiful tray of cupcakes, and off you go!

14.3 Informal Meetings for 20–50 People, for a Day or Less (Where by 'Informal', We Mean That You Have No Budget …)

Larger meetings entailing external speakers can still be very simple to organise providing you don't have to supply catering (which costs money, so you can't do this unless you have some money) and you don't have to worry about accommodation for people or their travel costs. People will come to such meetings and pay their own way, if the meetings are really great meetings and everyone mingles and talks about their science. And if the venue is nice and it's easy to get something to eat and drink. However, you have to be absolutely clear in all your communications with attendees and with speakers, that there's no budget for anything — travel, food, accommodation, T-shirts, merchandising, plastic chromosomes or toy atomic piles. If it's a good meeting, perhaps with a bunch of collaborators locally, then people will still pile in. Scientists are like that (generally).

So here is our step by step guide to organising a meeting with no budget.

Teamwork

Meetings can be hard work to organise and it's usually beyond the scope of one person, who may also be doing a PhD or trying to tackle a postdoctoral project. So get together a small team, between 3 and 5 people, to organise the meeting. Go with people who will step up and take the time to do what they say they'll do.

Subject

Decide on the core subject of the meeting and then come up with a title for it that is explicit and informative.

Date

Decide on a date that is realistic. Too close and no one will be able to make it. Too far away and everyone will have forgotten about it by the time the meeting comes around, 5 years after you thought of it. For a meeting of 20–50 people including super-busy lab heads, then aim to have the meeting 6–9 months in the future.

One of us organised a conference at a venue well away from home but forgot to check the date carefully. It turned out that the date chosen clashed with a particularly significant family occasion, which should have been obvious immediately. This was not good for family relations, and could easily have been avoided by checking the date in advance. This mistake is still frequently alluded to, several years after the event!

The most important speakers

If you have a few speakers who are absolutely essential for the success of your meeting, check immediately that they are willing to come and are available on the date you've chosen. If they're not willing to come, perhaps they could suggest someone from the same lab to present their work or similar data. If they're not available on your chosen date, you may need to change your date! Contact your speakers immediately because scientists are VERY busy people and often have dates blocked out in their diaries over a year in advance.

Incidentally, if you're organising the meeting it's also fine to have you as a speaker, but obviously keep a sense of proportion — don't schedule a 1 hour talk for yourself and a 5-minute talk for the Fields Medal winner who has graciously agreed to attend.

The venue

First of all book the room — but you have no budget so the room has to be free for you. Your chosen room also has to be big enough for your meeting and to have the projection facilities you need. And, as we mentioned above, go for the nicest room you can, especially if you have to spend all day in it.

The timetable

Let's start with the talks. How long should the talks be? They don't all have to be the same length, and you need to include 5–10 minutes for questions. As a general rule no one is capable of listening beyond 40 minutes, so make your longest talks, which will have the most questions, 40 + 10 = 50 minutes long, i.e. a 40-minute talk with 10 minutes for questions afterwards. If you want to maximise the number of speakers, on the assumption that people get a taste of what they're doing and can ask more detailed questions afterwards, you can always ask people to talk for 15 + 5, or even 12 + 3 minutes if you're really short of time.

So now you have one or two main speakers, you've decided about the mix of talks, and the next step is you need to create the timetable in order to work out your total speaker number. This means you also need to decide how long the breaks will be. If you're not providing catering, then people have to have time to get to a nearby source of coffee/tea/water, and get back again. And go to the toilet (which must be easily visible or signed on the day, including disabled facilities).

Also, if you have a session of more than 2 hours, we guarantee at least 25% of your audience will be asleep. *Never* go beyond 2 hours without a break, and try to stick to a maximum of 1.5 hours for each session of talks.

So if you've sorted out all that, what's the length of the meeting? People need time to arrive and if you're not providing a dinner they need time to get home.

OK, so *finally* you've decided all that, let's draw up the timetable. This often takes a few drafts, even for small half-day meetings. For example, here's a typical timetable for a 1-day meeting without a budget. The meeting starts at 10am so that people have time to arrive, and it finishes at 5pm. Which means that you have to be there from at least 9.30am to meet and greet — and if you want to give people a printed timetable, or just say hello, you can do so.

09.00	You arrive, make sure the room is unlocked, the toilets are signed and the projector is working.
09.30	Your speakers and attendees start to show up. The attendees want to know where to buy coffee.
10.00–10.05	You introduce the meeting. State the topic, tell people where the toilets are and the catering. Thank whoever has helped you organise the meeting. If anyone has.
10.05–10.55	Important speaker talking for 40 minutes plus 10 minutes for questions.

10.55–11.15	Speaker giving 15 minute talk with 5 minutes for questions.
11.15–11.45	Half hour break, which gives enough time for everyone to pile into a nearby coffee shop for refreshments.
11.45–12.05	Speaker giving 15 minute talk with 5 minutes for questions.
12.05–12.25	Speaker giving 15 minute talk with 5 minutes for questions.
12.25–12.45	Speaker giving 15 minute talk with 5 minutes for questions.
12.45–14.00	Everyone splits for lunch. You need to tell them where to go. Do they have enough time to queue for a cheese and tomato sandwich and get back to the venue? It's good to be generous with the time for lunch so people have a good chance to talk to each other as well, and just in case the morning session over runs a little.
14.00–14.35	Speaker talking for 30 minutes plus 5 minutes for questions.
14.35–15.10	Speaker talking for 30 minutes plus 5 minutes for questions.
15.10–15.40	Everyone is ready for another break by now. They need 30 minutes.
15.40–16.00	Speaker talking for 15 minutes plus 5 minutes for questions.
16.00–16.20	Speaker talking for 15 minutes plus 5 minutes for questions.
16.20–16.40	Speaker talking for 15 minutes plus 5 minutes for questions.
16.40–17.00	Speaker talking for 15 minutes plus 5 minutes for questions.
17.00–17.01	You close the meeting by thanking everyone for attending.
17.02	You lock the door of the venue, return the key to whoever needs it, turn out the lights, retire, exhausted but happy and successful, to the pub …

So that's a typical timetable, and note two things (1) do not go past 5pm as some people may have childcare and other duties after that and (2) your timetable constrains the number of speakers you can have. Even with *lots* of short talks, the speaker number above is only (count them): 11 speakers.

The other speakers

You have one or two key people already, you know roughly how many other speakers you can fit in, you know whether you want long or short talks, and so now you can decide who else to invite, and it's often also a good idea to keep one or two slots free in case good speakers or people with hot new data ask if they can talk.

Think of who is a good speaker, who has interesting data, who comes from an important lab, who is enthusiastic. You may not get all of that in one person, but at least aim for a mix of those features throughout the day. Take advice from colleagues to help get a good selection of people to speak. Get your list of names — and make sure you have a good balance of men and women, speakers from your own country and elsewhere — and contact your chosen speakers immediately. If it's your friend along the corridor a short conversation may be enough of an invitation, but follow it up with a reminder email telling them exactly when and where the meeting is, for their records and their diary. This also means next time you see them you can check if they really have put the meeting in their diary and won't double book themselves.

Tell each speaker the rough area of research that you would like them to present and be completely clear that you have no money for a budget and that means no money for catering or for travel expenses.

If you don't know your speaker at all then write a polite email explaining the topic of the meeting, when and where it is, perhaps who else is speaking, and what you'd like them to talk about. Again, make it clear there is no budget for expenses, or catering.

If they're a scarily famous big cheese with a legendarily bad temper, then write a *very* polite email. Here is a letter sent to Professor Karloff of the Department of Parapsychology, University of West

Cheam, inviting him to a conference on Experimental Levitation, within the Department of Paranormal Engineering, University of West Cheam. This letter gives him all the practical details he needs (where and when), lets him know that he will have to pay his own way (fortunately the two Departments are in adjacent buildings) and it also shows him the size of the meeting and that other important people in his field are being invited.

Dear Professor Karloff,

We are writing to invite you to give a talk at the conference we are organising on the topic of Experimental Levitation, on Tuesday 21st July 2020. This is a one-day meeting that will take place in the Mary Shelley Lecture Theatre, Department of Paranormal Engineering, University of West Cheam.

We would be delighted if you were able to talk on the topic of the Parapsychology of Landfish. Other invited speakers include Professors Hazel Witch FRS and Frank N. Stein. We do not have funding for this meeting which is an informal workshop of approximately fifty attendees, and so we are not able to pay travel or other expenses.

We would be honoured if you were able to accept our invitation, and we look forward to hearing from you.

Best wishes,

The organisers

Professor Karloff also received this invitation.

```
Hi Barold,

We're organising a meeting and want yu to talk about
your research. It's in July 2020 in our Dept and we
want you to talk about Landfish.

Thanks for saying yes!!!

The organisers.
```

Well, which invitation do you think the irascible Professor Karloff accepted, and which did he throw in a rage into his email trash?

Publicising your meeting and thinking about audience number

You have your speakers, you have your timetable and your venue. What about your audience? If it's a small local meeting you probably have a list of people to email, and that is the best way to tell people of the meeting. Organisers of earlier conferences in your field may also have an email list that you could use. You may be able to gauge how many people will turn up, given how popular your field is and how many people you tell about the meeting.

However, you may want to make things more open and perhaps have your Departmental or University website advertise the meeting, or even your Learned Society or Professional Body. In that case, think about what to do if you get a lot more or less interest than expected. If you're worried about numbers and have the chance to switch to a smaller or a larger venue then do you want people to register beforehand so that you have a rough idea of numbers? Or is it fine if they just turn up on the day?

Whatever you decide, keep sending out emails. People forget and then even if they haven't forgotten, they have accidentally deleted the email with the venue details. So make it easy for people to know about, and to come to, your meeting.

Cash, money, moolah, spondulix, dough, gold, dosh

You may decide you want to charge people a small nominal fee for coffee and perhaps a glass of wine afterwards, to facilitate networking. That is a completely legitimate and important aim. If you do this, you have to be prepared for the worst — too little money to cover the costs — and you have to be super organised about where the money goes and how you collect it. Unless you have your own credit card reader, cash on the day is easiest, and it's also the easiest item to lose. Think responsibly about how to collect money, talk to an Administrator at work, and if you are going to charge, make sure people know in advance that they have to pay, how much it is, and that you only accept cash. Regardless of how much you publicise this, 10% of your audience will turn up with no money. So make sure you know where the nearest bank/automatic teller is.

More essentials

Ensure any local rules are followed, that people know where the toilets are (look, honestly, this is important, which is why we keep saying it) and the fire exits. And the coffee source.

Ask around for money to help out with small expenses, or even for coffee. Ask your lab head (who will say 'no'), ask your Department, your Administrator, your Head of Department, your Dean, ask anyone you can think of. You could try a publisher or equipment supplier in your field as well. They may be happy to provide some cash in exchange for some sort of visibility, e.g. leaflets or a table to display their wares. Be polite and be reasonable and sometimes you can be pleasantly surprised. Don't get down hearted if you have no budget. Some of the best meetings are the small informal ones, where people are relaxed, chipping in, and making friends.

Final preparations

Before the day: send a reminder to speakers a week beforehand. Include your mobile number in case they get into difficulties on the day. People get involved in car crashes, their trains break down, the bus gets lost, and you need to know where your speakers are. And, very important: TELL THE SPEAKERS WHAT THEY NEED TO BRING FOR THEIR PRESENTATION. A LAPTOP? OR IS A USB STICK OK? If you want to use one laptop for all the presentations, get the speakers to email them to you in advance (but remember that some of them probably won't …).

On the day: come early to meet and greet. Think about what the speakers might need (a laser pointer?) and bring it along.

Not all your audience will turn up. Don't worry about this. We reckon that for any meeting about 10% of people who've said they want to attend won't turn up. That's just how it is.

If a speaker can't make it at the last minute don't panic. It's more time for questions or for lunch. Just say to the audience that the speaker has been unable to attend at the last minute, you don't need to give a reason, and shuffle the timetable in the way that works best. Perhaps you can have a longer lunch break if it's glorious weather and people are clearly enjoying chatting on the lawns and listening to the peacocks, or perhaps it's starting to snow and everyone will appreciate finishing earlier.

One of us went to a workshop organised by a group of PhD students. When they announced the arrangements for the conference dinner they suggested that everyone just got the bus from the University campus into the city centre. No-one had thought that 50 additional passengers were very unlikely to all fit on the bus!

14.4 Larger Meetings Where You Need a Budget

As you can see, it's a lot of work just organising a small meeting with no budget. And sadly, what you can do largely depends on your budget.

Ah money, money. It makes the world go around. It also enables us to have meetings with expenses for travel costs and for catering and even Abstract booklets. So if you are thinking of holding a meeting you need to find out what funds are available to support your meeting. This may entail writing a proper grant application, or having a chat with your wealthy Head of Department, or (as we know only too well) writing begging letters to Industry. Whatever the process, you need to work well in advance and to know what the topic of your meeting is, how long it is on for, and who you want to talk in order to attract in an interested audience.

It's a HUGE amount of work organising a big meeting (we've been there, done it, and possibly regretted it). That's why most large meetings nowadays are organised by professional conference organisers (which itself increases costs of course). They have the contacts and expertise needed to make conferences run smoothly. Many Learned Societies offer this sort of service professionally. So for large meetings our advice is to leave it to the experts to do the work!

Key Points

- Don't underestimate how much time it will take you to organise a meeting!
- If you can, get a budget! It will make everything else much easier if you have some money.
- Get other colleagues to help you organise the meeting if you can.
- Look for a suitable venue with the facilities you need.
- Good catering arrangements help everything run smoothly so organise catering carefully.
- Be realistic with your timetable and don't try to cram too many talks into the time available.
- Contact potential speakers and attendees in good time.

Chapter 15

Conclusion

Scientific conferences have been taking place for many many years because people have found them to be an effective way of communicating. Nowadays it would be possible to use electronic means such as Skype to keep in touch with people, and many talks are available on YouTube. But despite this, people still want to get together in person to talk and discuss because the quality of face to face interaction is so much better.

As we have discussed in this book, conferences are really important for the progress of science. They give the opportunity for you to tell other people about your research — your methods, your ideas, your theories and your results. And similarly they allow you to find out about other people's research. Conferences are good for discovering the latest advances — books and papers tend to come much later.

Conferences are also good for talking to people and cementing professional and personal relationships. That gives a firm foundation for later interactions after the conference. One important result of this may be job opportunities for you elsewhere or for other people in your own group.

In order to make the best of all these opportunities, you have to make good preparations, as we have discussed in this book. First, you need to do your research to find the best conference for you. Then you need to write a good Abstract that sums up what you want to communicate at the conference, bearing in mind that this may be necessary in order to be invited to attend and that the Abstract will be a permanent record of your contribution after the conference. Most importantly, you need to prepare your talk or poster with great care so that your communication at the conference is effective. Everyone who has been to a conference knows that it is the well-prepared talks and posters that are the most enjoyable and useful.

Similarly, at the conference it's important to use your time wisely and make the most of the chances to hear people speak and to talk to them at their poster or at other times at the conference.

Finally, we all benefit from other people's investment of time and energy into organising conferences and so it is only reasonable that we put something back by being prepared to organise conferences ourselves.

We hope that the advice and suggestions we have set out in this book will help you not to be intimidated by the process of preparing for and attending a conference, and that therefore you will be able to make good use of the opportunities offered by a conference and benefit from attending. Above all, we hope that the experience will be a positive one and that you will enjoy attending your scientific conference!

Index